T0265336

Advances in Computer Vision and Pattern Recognition

Founding editor

Sameer Singh, Rail Vision, Castle Donington, UK

Series editor

Sing Bing Kang, Microsoft Research, Redmond, WA, USA

More information about this series at http://www.springer.com/series/4205

Ajita Rattani · Fabio Roli · Eric Granger
Editors

Adaptive Biometric Systems

Recent Advances and Challenges

 Springer

Editors
Ajita Rattani
Department of Computer Science
  and Engineering
Michigan State University
East Lansing, MI
USA

Eric Granger
Department of Automated Manufacturing
  Engineering
École de technologie supérieure
Montreal, QC
Canada

Fabio Roli
Department of Electrical and Electronic
  Engineering
University of Cagliari
Cagliari
Italy

ISSN 2191-6586   ISSN 2191-6594   (electronic)
Advances in Computer Vision and Pattern Recognition
ISBN 978-3-319-24863-9   ISBN 978-3-319-24865-3   (eBook)
DOI 10.1007/978-3-319-24865-3

Library of Congress Control Number: 2015950048

Springer Cham Heidelberg New York Dordrecht London

Printed on acid-free paper

Springer International Publishing AG Switzerland is part of Springer Science+Business Media
(www.springer.com)

Preface

Biometrics is the science of recognizing individuals on the basis of their physical (such as face, fingerprint, and iris) or behavioural traits (such as voice and gait). It holds a lot of promise over traditional password-based systems, such as PIN and password, therefore revolutionizing the way authentication is done. Biometric applications include border crossing, national civil registry, smartphone security, mobile payment and access to restricted facilities.

Continual improvements in accuracy, transaction speed, affordability of biometric systems and technologies have increased their ease of use and cost-effectiveness. While biometric technology continues to be adopted, an intrinsic characteristic of the technology is that system error rate simply cannot attain absolute zero in real-world applications. The main cause for mismatch errors is the variable acquisition conditions in semi- and uncontrolled environments, due to changes in pose, illumination, human–sensor interactions, occlusions, expressions, ageing, etc.

In addition to complex operational environments that change over time, biometric systems are typically designed a priori using limited and unbalanced data and without any knowledge of underlying data distributions. Therefore, biometric models may be often poor representatives of the biometric trait to be recognized, and should be adapted over time in response to new or changing input features, quality of the input data samples, change in sensor/matching algorithm and environments. Several innovative techniques have recently emerged to adapt the biometric system over time. These systems are collectively termed as *adaptive biometric systems*.

Recently, adaptive biometrics has gained much attention from the research community, and is expected to continue this momentum because of its key promulgated features. First, with this system, one no longer needs to collect a large number of biometric samples during enrollment. Second, it is no longer necessary to re-enroll or re-design the biometric system (classifier) from scratch in order to cope up with changing environments. This convenience can significantly reduce the cost of maintaining a biometric system. Third, the actual observed intraclass variations like aging can be incorporated into the system. In fact, biometric vendors such as BIOsingle (fingerprint) and Recogsys (hand geometry) have incorporated the automated adaptation mechanism into their technologies. However, there are many challenges and research issues to be solved such as the possibility of corruption of biometric models or template galleries with impostor intrusion due to the overlap in their respective genuine and impostor score distribution, the informative patterns, the stopping criteria of adaptive biometric system, etc.

Overall, this book aims to present a clear understanding of, recent advances and challenges to promote the field of adaptive biometric systems. Further, this book is a collection of numerous techniques to biometric system adaptation under unified taxonomy. Furthermore, adaptation procedures specified in this field are applicable to any pattern recognition system. This book is suitable for final-year undergraduate students, postgraduate students, engineers, researchers and academicians in the field of computer engineering who are engaged in various disciplines of system sciences, information security and identity businesses. We are indebted to a number of individuals in academic circles who have contributed in different, but important, ways to the preparation of this book. In particular, we would like to extend our appreciation to Arun Ross, Walter Scheirer, Reza Derakhshani, Massimo Tistarelli, Phalguni Gupta, Aurobindo Chatterjee, Gian Luca Marcialis, Norman Poh, Vinay Budhraja, Vijeta Rattani, Zahid Akhtar, Hunny Mehrotra, Davide Ariu, Biagio Freni and Ruggero Donida Labati. The objective of this book is also to engage researchers from academia and industry on the state-of-the-art biometric research and technology, and the potential problems in real applications.

Contents

Contributors

Amr Ahmed University of Lincoln, Lincoln, UK

Zahid Akhtar University of Udine, Udine, Italy

Selma Belgacem University of Rouen, Saint-etienne du Rouvray, France

M. Blumenstein Institute for Integrated and Intelligent Systems, Griffith University, Queensland, Australia

Clement Chatelain University of Rouen, Saint-etienne du Rouvray, France

A. Das Institute for Integrated and Intelligent Systems, Griffith University, Queensland, Australia

Cigdem Eroglu Erdem Bahcesehir University, Istanbul, Turkey

M.A. Ferrer IDeTIC, University of Las Palmas de Gran Canaria, Las Palmas, Spain

Gian Luca Foresti University of Udine, Udine, Italy

E. Granger Lab. d'Imagerie, de Vision et d'Intelligence Artificielle, École de Technologie Supérieure, Université du Québec, Montreal, Canada

Phalguni Gupta National Institute of Technical Teachers' Training and Research, Kolkata, India

Joseph Kittler University of Surrey, Surrey, UK

R. Kunwar Institute for Integrated and Intelligent Systems, Griffith University, Queensland, Australia

G.L. Marcialis Pattern Recognition and Applications Group, Department of Electrical and Electronic Engineering, University of Cagliari, Cagliari, Italy

C. Pagano Lab. d'Imagerie, de Vision et d'Intelligence Artificielle, École de Technologie Supérieure, Université du Québec, Montreal, Canada

U. Pal Computer Vision and Pattern Recognition Unit, Indian Statistical Institute, Kolkata, India

Thierry Paquet University of Rouen, Saint-etienne du Rouvray, France

Norman Poh University of Surrey, Surrey, UK

Ajita Rattani Michigan State University, East Lansing, MI, USA

F. Roli Pattern Recognition and Applications Group, Department of Electrical and Electronic Engineering, University of Cagliari, Cagliari, Italy

R. Sabourin Lab. d'Imagerie, de Vision et d'Intelligence Artificielle, École de Technologie Supérieure, Université du Québec, Montreal, Canada

Kamlesh Tiwari Indian Institute of Technology Kanpur, Kanpur, India

P. Tuveri Pattern Recognition and Applications Group, Department of Electrical and Electronic Engineering, University of Cagliari, Cagliari, Italy

Chapter 1
Introduction to Adaptive Biometric Systems

Ajita Rattani

Abstract Biometric person recognition poses a very challenging pattern recognition problem because of large variability in biometric sample quality encountered during testing and a restricted number of enrollment samples for training. Furthermore, biometric traits can change over time due to aging and change of lifestyle. Effectively, the noise factors encountered in testing cannot be represented by the limited training samples. A promising solution to training data deficiency and ageing is to use an *adaptive biometric system*. These systems attempt to adapt themselves to follow the change in the input biometric data. Adaptive biometrics deserves a treatment on its own right because standard machine-learning algorithms cannot readily handle changing signal quality. The aim of this chapter is to introduce the concept of adaptive biometric systems in terms of taxonomy, level of adaptation, open issues and challenges involved.

1.1 Introduction

While the biometric technology continues to improve, an intrinsic characteristic of the technology is that the system error rate simply cannot attain the absolute zero [1, 2]. The major cause for biometric recognition errors is the compound effect of the inherent scarcity of training samples during the enrollment phase as well as the presence of substantial sample variations during the operational phase. The large sample variation is caused by the vulnerable nature of data acquisition process and the external changing acquisition conditions [1, 3–5]. Moreover, the system is also expected to match biometric samples acquired by different devices that could give significantly different levels of quality [6]. Apart from this, being biological tissues in nature, biometric traits can be altered either temporarily or permanently, due to ageing [7–10], diseases, or treatment to diseases [3, 4].

A. Rattani (✉)
Michigan State University, East Lansing, USA
e-mail: ajita@msu.edu

© Springer International Publishing Switzerland 2015
A. Rattani et al. (eds.), *Adaptive Biometric Systems*,
Advances in Computer Vision and Pattern Recognition,
DOI 10.1007/978-3-319-24865-3_1

With all the above adverse factors, a biometric system cannot be expected to maintain its performance over a long period of time and cope up with all possible sources of variation. A number of solutions have been proposed to reduce the impact of sample variation, namely multi-biometrics, feature invariance and signal restoration schemes, use of computer graphic techniques to simulate age progression/regression, and recently, adaptive biometric systems.

Although *multibiometrics* [11] can improve the robustness of a biometric system, this solution cannot account for genuine changes in biometric traits due to the lapse of time or ageing. Feature invariance and signal restoration schemes aim to render biometric feature robust to noise. For instance, in face recognition, a number of techniques have been developed to normalize face images against lighting via illumination normalization [12], against pose or expression change via affine transformation, and against the presence of glasses by glass-removal algorithms. However, due to the large number of possible sources of variations, attempting to normalize against one factor at a time could introduce artefacts. This is most notable for extreme pose correction. However, more importantly, the procedure does not account for biometric trait change due to ageing and lifestyle-related changes.

Computer graphic techniques have also been used to synthesize a novel view of a given biometric sample. For instance, invariance to facial ageing can be achieved by simulating the effect of ageing in personal faces [3, 6, 13, 14]. Although photo-realistic images can be simulated, these methods by no means take into account the complex underlying ageing process, which is a function of a person's lifestyle and health status. Furthermore, the synthesis process often relies on an initialization process that can sometimes be prone to estimation error if not manually corrected.

Recently, adaptive biometrics have been introduced as a solution to track the changes as well as to learn the intra-class variation of biometric samples by allowing the underlying *biometric reference*, which can be a template or a model,[1] to be updated using *operational data*. Unlike a traditional system, an adaptive biometric system has an additional module called *adaptation* or updating module. The aim of this module is to continuously adapt the system to the intra-class variation of the input data as a result of (1) changing acquisition conditions and (2) age and lifestyle-related changes. These two strategies of adaptation are called *condition* and *age* adaptation, respectively.

As the update process is invoked multiple times, several references will be generated. This requires a *reference management* strategy. One strategy is to maintain only a single large common reference that embeds all the information [14]. This is called the *super-template* approach. Another strategy is to update the existing reference by replacing or appending the newly acquired input sample to the template set. This strategy is commonly used for model-based reference; it involves re-estimating or updating the model parameters.

---

[1] A template refers to the biometric sample used for enrollment. The term "model" refers to statistical representation derived from one or more biometric samples. In order for our discussion to cover both types of methods, we shall adapt the standard vocabulary, that is, "biometric reference" or simply reference. A reference is subsequently used for comparing a biometric test/query sample to obtain a similarity score.

Among the above solutions, adaptive biometrics is arguably the most recent solution being explored. The unique advantage of this approach is that the actual observed variation can be incorporated into the reference. Nevertheless, adaptive biometrics has high future prospects because of its key promulgated advantages. First, with an adaptive biometric system, one no longer needs to collect a large number of biometric samples during enrollment. Second, it is no longer necessary to re-enrol or re-train the system (classifier) from scratch in order to cope up with the changing environment [6].

The aim of this chapter is to introduce the concept of adaptive biometric systems in terms of taxonomy, level of adaptation, open issues and challenges involved. Next, we discuss the attributes of existing adaptive biometric systems.

1.2 Attributes of the Existing Adaptive Biometric systems

In an attempt to categorize adaptive biometric systems, the most logical way to proceed is to define a number of key attributes [3, 4, 15]. These attributes are explained below:

1. *Supervised versus Unsupervised*: In the supervised adaptation scenario, a (human) supervisor is available to label the input data [11]. In contrast, in an unsupervised scenario, the label of the data (whether being a match or not) is unknown to the system. The system attempts to infer the label and only those samples whose labels can be inferred confidently are used to adapt the reference [6]. It is expected that supervised adaptation always results in the most optimistic performance, that is, the best achievable performance compared to an unsupervised one. Figures 1.1 and 1.2 illustrate an example of supervised (input samples are labeled by the supervisor) and unsupervised learning (where the input samples are automatically labeled by the biometric system).

2. *Static versus video based*: The type of data used for adaptation can also make a difference. In particular, in a video-based biometrics, one can exploit the fact that the person whose biometrics is being sampled remains the same for the entire video sequence, since each consecutive pair of images in the sequence are a fraction of the second apart from each other. This is the *identity constancy* property [4, 14]. In comparison, one cannot exploit this property from static images that are obtained from a biometric sensor because only one image is sampled in a single acquisition session. Consequently, this is a harder problem as the identity constancy property cannot be exploited. This manuscript focuses on the static image-based adaptive biometric systems.

3. *Level of adaptation*: In addition to the adaptation at the reference (template or model) level, the process of adaptation can also take place at the score or decision level. In [16], biometric sample quality is used to adapt the matching score so as to render the final accept/reject decision independent of the sample quality. The rationale behind adaptive score-level normalization is that the distributions of

Fig. 1.1 Illustration of supervised adaptation scheme in which the input samples are labeled by the supervisor. The positively labeled samples are used to adapt the biometric system, taken from [6]

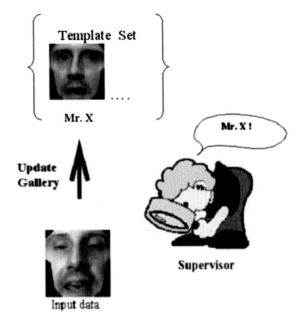

match (genuine) and non-match (impostor) score are dependent on the condition of acquisition. Adaptation at the decision level adapts the decision module to the changing conditions.

4. *Self- versus co-training*: In the case of semi-supervised adaptation, the system will need to infer the label from the operational data. Two commonly used strategies are self-training and co-training [17]. In *self-training*, the algorithm uses highly confidently classified input samples to update the reference [3, 4, 18–21]. An offline self-training procedure may also use a label propagation scheme [22] to determine whether or not the operational samples should be used for adaptation. In *Co-training*, the mutual and complementary help of two biometrics is used to adapt the reference. According to [15, 18], co-training can better capture input samples with much more significant variation, hence, resulting in better performance than self-training.

5. *Online versus offline adaptation*: The availability of computer memory can also determine the type of adaptation. When the memory size is limited, one resolves to employing an online adaptation strategy in which case the system updates its parameter as soon as an input sample has been successfully authenticated and deemed suitable for adaptation (online) [20, 23]. When a large buffer or memory is available, the adaptation process can be delayed until a later point in time or when the buffer is full. We refer to this adaptation strategy as *offline* [15, 21, 22].

6. *Quality versus non-quality based adaptation*: Recent advancement in the biometric community shows that quality has considerable impact on the system performance for various traits like fingerprint, iris, face, etc. [14]. Quality measures

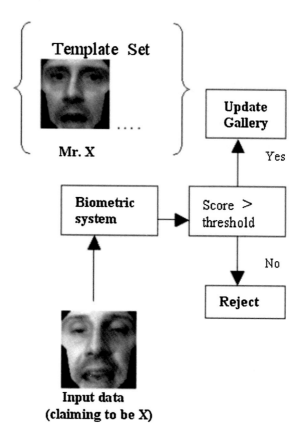

are an array of measurements quantifying the degree of excellence or confor-
mance of biometric samples to some predefined criteria known to influence the
system performance. However, it is only recently that biometric sample quality has been considered for adaptive biometric systems [15, 18]. Quality-based
adaptation requires maintaining a different set of updated models for each type
of condition. Since a query sample is always acquired under a particular condition, the inference (matching task) requires the identification of the condition.
The inference problem can be formulated using a Bayesian framework [22]. In
this manuscript, the resultant system is called a *condition-adaptive* system.

1.3 Open Issues and New Research Directions

Adaptive biometrics is a challenging topic. Although template update methods in
adaptive biometrics have shown to be promising, some open issues still need to be
addressed for their effective implementation. In particular, the existence, if any, of
the tradeoff between performance enhancement and gallery size is maintained due to

updating. Worth mentioning, all the template update methods are prone to impostors introduction and the attraction of more samples of it may gradually lead to creep-in of identity, when the genuine person loses its identity. Figure 1.3 shows an example of mis-classification error leading to creep-in of identity problem.

Studies [3, 24] have reported that even with the operation of update procedures at stringent threshold condition, the introduction of impostors cannot be avoidable. As apart from factors like incorrect estimation of threshold or basic FAR of the system, these methods are much prone to impostors introduction due to the presence of difficult clients, wolves and lambs, according to the Doddingtons zoo [24]. Wolves are clients having the ability to imitate others irrespective of stringent threshold conditions while lambs are clients vulnerable to impostors attack and the presence of these characteristic clients result in impostor introduction. To model the early stoppage of impostor introduction due to these client is still an open issue [4].

Existing studies suggest that mis-classification errors result in degradation in the performance of adaptive biometric systems. In other words, adaptive biometric systems considering impostor attacks result in lower performance gain in comparison to those using only genuine samples for adaptation. This is on account of updating using impostors as a result of intrinsic failure of the system, i.e., false accept rate (FAR); thus increasing the vulnerability to template security and undermining the integrity of the adaptive biometric systems. To this front, modelling and early stoppage of impostor attack into the updated template set is an important research direction to be pursued. Avoiding impostor intrusion into the updated template set will allow commercial vendors to adopt auto-update procedures in their commercial biometric products.

Further, there is a need for a *robust adaptation scheme* incorporating optimum labelling procedure for the input samples. This is supported by our findings related to supervised versus semi-supervised methods for adaptation where the supervised scheme generalizes better than the semi-supervised one. This indicates that the use of confidently classified input samples (as used by most of the existing automated systems based on semi-supervised learning) may not be an efficient strategy for

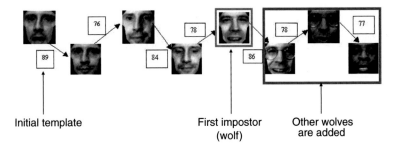

Initial template   First impostor   Other wolves
                                              (wolf)   are added

Fig. 1.3 Mis-classification error by the adaptive biometric system, leading to creep-in of identity problem, taken from [24]

adaptation. These results emphasize the *need for more robust adaptation schemes* that are capable of identifying genuine samples with substantial variations without increasing the vulnerability to impostor intrusion [3, 4, 15].

1.4 Conclusion

In this chapter, we introduced the concept of adaptive biometrics, discussed the taxonomy for summarizing the current state of the art and highlighted the open issues and challenges involved. Although adaptive biometric systems have shown to be a promising solution, some open issues still need to be addressed for their effective implementation. In particular, the existence of the tradeoff between performance enhancement and gallery size maintained due to updating. Worth mentioning, all the adaptive methods are prone to impostors introduction and the attraction of more samples of it may gradually lead to creep-in of identity, when the genuine person loses its identity. The problem of impostor introduction has been stated in [15]. Reference [24] has reported that even with operation of update procedures at stringent threshold condition, the introduction of impostor cannot be avoidable. As apart from factors like incorrect estimation of threshold or basic FAR of the system, these methods are much prone to impostors introduction due to the presence of biometric menagerie, i.e., wolves and lambs according to the Doddington zoo [24] concept. Wolves are users having the ability to imitate others irrespective of stringent threshold conditions while lambs are users vulnerable to impostors attack and the presence of these characteristic users result in impostor introduction. To model the early stoppage of impostor introduction due to these difficult users is still an open issue. The adaptive methods introduced in this chapter and the open issues discussed, are applicable to any adaptive pattern recognition system in general.

References

1. Akhtar, Z., Ahmed, A., Erdem, C.E., Foresti, G.L.: Biometric template update under facial aging. In: Proceedings of the IEEE Symposium on Computational Intelligence in Biometrics and Identity Management (IEEE SSCI-CIBIM) (2014)
2. Poh, N., Kittler, J., Rattani, A., Tistarelli, M.: Group-specific score normalization for biometric systems. In: Proceedings of the IEEE Conference on Computer Vision and Pattern Recognition Workshops (CVPRW), pp. 38–45 (2010)
3. Rattani, A., Marcialis, G.L., Roli, F.: Biometric system adaptation by self-update and graph-based techniques. J. Vis. Lang. Comput. **24**(1), 1–9 (2013)
4. Rattani, A., Marcialis, G.L., Granger, E., Roli, F.: A dual-staged classification-selection approach for automated update of biometric templates. In: International Conference on Pattern Recognition (ICPR), pp. 2972–2975 (2014)
5. Tsymbal, A.: The problem of concept drift: definitions and related work. Department of Computer Science, Trinity College, Ireland (2004)

6. Rattani, A.: Adaptive biometric system based on template update procedures, Ph.D. thesis, University of Cagliari, Italy (2010)
7. Akhtar, Z., Rattani, A., Hadid, A., Tistarelli, M.: Face recognition under ageing effect: a comparative analysis. In: Proceedings of the International Conference on Image Analysis and Processing (ICIAP), pp. 309–318 (2013)
8. Ling, H., Soatto, S., Ramanathan, N., Jacobs, D.W.: A study of face recognition as people age. In: Proceedings of the 11th IEEE International Conference on Computer Vision (ICCV), pp. 1–8 (2007)
9. Ramanathan, N., Chellappa, R.: Face verification across age progression. In: Proceedings of the IEEE Conference Computer Vision and Pattern Recognition (CVPR), pp. 462–469 (2005)
10. Lanitis, A., Taylor, C.J., Cootes, T.F.: Toward automatic simulation of aging effects on face images. IEEE Tran. Pattern Anal. Mach. Intell. **24**(4), 442–455 (2002)
11. Uludag, U., Ross, A., Jain, A.: Biometric template selection and update: a case study in fingerprints. Pattern Recognit. **37**(7), 1533–1542 (2004)
12. Ahonen, T., Hadid, A., Pietikainen, M.: Face description with local binary patterns: application to face recognition. IEEE Trans. Pattern Anal. Mach. Intell. **28**(12), 2037–2041 (2006)
13. Rattani, A., Marcialis, G.L., Roli, F.: Temporal analysis of biometric template update procedures in uncontrolled environment. In: Proceedings of 16th International Conference on Image Analysis and Processing (ICIAP'11), Raveena, Italy, pp. 595–604 (2011)
14. Poh, N., Rattani, A., Roli, F.: Critical analysis of adaptive biometric systems. IET Biom. **1**(4), 179–187 (2012)
15. Rattani, A., Freni, B., Marcialis, G.L., Roli, F.: Template update methods in adaptive biometric systems: a critical review. In: Proceedings of the International Conference on Biometrics (ICB), pp. 847–857 (2009)
16. Pavani, S.K., Sukno, F.M., Butakoff, C., Planes, X., Frangi, A.F.: A confidence based update rule for self-updating human face recognition systems. In: Proceedings of the International Conference on Biometrics (ICB), pp. 151–160 (2009)
17. Rattani, A., Marcialis, G.L., Roli, F.: Biometric template update using the graph mincut: a case study in face verification. In: Proceedings of the 6th IEEE Biometric Symposium (2008)
18. Liu, X., Chen, T., Thornton, S.M.: Eigenspace updating for non-stationary process and its application to face recognition. Pattern Recognit. 1945–1959 (2003)
19. Franco, A., Maio, D., Maltoni, D.: Incremental template updating for face recognition in home environments. Pattern Recognit. **43**, 2891–2903 (2010)
20. Jiang, X., Ser, W.: Online fingerprint template improvement. IEEE Tran. PAMI **8**, 1121–1126 (2002)
21. Roli, F., Marcialis, G.L.: Semi-supervised PCA-based face recognition using self training. In: Proceedings of the Joint ICPR International Workshop on S+SSPR (2006)
22. Poh, N., Kittler, J., Marcel, S., Matrouf, D., Bonastre, J.F.: Model and score adaptation for biometric systems: coping with device interoperability and changing acquisition conditions. In: Proceedings of the 20th International Conference on Pattern Recognition (ICPR), pp. 1229–1232 (2010)
23. Ryu, C., Hakil, K., Jain, A.: Template adaptation based fingerprint verification. In: Proceedings of the International Conference on Pattern Recognition (ICPR), pp. 582–585 (2006)
24. Rattani, A., Marcialis, G.L., Roli, F.: An experimental analysis of the relationship between biometric template update and the doddingtons zoo in face verification. In: Proceedings of 14th International Conference on Image Analysis and Processing (ICIAP'09), Vietri sul Mare (Italy), pp. 434–442 (2009)

Chapter 2
Context-Sensitive Self-Updating for Adaptive Face Recognition

C. Pagano, E. Granger, R. Sabourin, P. Tuveri, G.L. Marcialis
and F. Roli

Abstract Performance of state-of-the-art face recognition (FR) systems is known to be significantly affected by variations in facial appearance, caused mainly by changes in capture conditions and physiology. While individuals are often enrolled to a FR system using a limited number of reference face captures, adapting facial models through re-enrollment or through self-updating with highly confident operational captures has been shown to maintain or improve performance. However, frequent re-enrollment and updating can become very costly, and facial models may be corrupted if misclassified face captures are used for self-updating. This chapter presents an overview of adaptive FR systems that perform self-updating of facial models using operational (unlabelled) data. Adaptive template matching systems are first revised, with a particular focus on system complexity control using template management techniques. A new *context-sensitive* self-updating approach is proposed to self-update only when highly confident operational data depict new capture conditions. This allows to enhance the modelling of intra-class variations, while mitigating the growth of the system by filtering out redundant information, thus reducing the need to use costly template management techniques during operations. A particular

C. Pagano (✉) · E. Granger · R. Sabourin
Lab. d'Imagerie, de Vision et d'Intelligence Artificielle, École de Technologie Supérieure,
Université du Québec, Montreal, Canada
e-mail: cpagano@livia.etsmtl.ca

E. Granger
e-mail: eric.granger@etsmtl.ca

R. Sabourin
e-mail: robert.sabourin@etsmtl.ca

P. Tuveri · G.-L. Marcialis · F. Roli
Pattern Recognition and Applications Group, Department of Electrical and Electronic
Engineering, University of Cagliari, Cagliari, Italy
e-mail: pierluigi.tuveri@diee.unica.it

G.-L. Marcialis
e-mail: marcialis@diee.unica.it

F. Roli
e-mail: roli@diee.unica.it

© Springer International Publishing Switzerland 2015
A. Rattani et al. (eds.), *Adaptive Biometric Systems*,
Advances in Computer Vision and Pattern Recognition,
DOI 10.1007/978-3-319-24865-3_2

9

implementation is proposed, where highly confident templates are added according to variations in illumination conditions detected using a global luminance distortion measures. Experimental results using three publicly available FR databases indicate that this approach enables to maintain a level of classification performance comparable to standard self-updating template matching systems, while significantly reducing the memory and computational complexity over time.

2.1 Introduction

Automated face recognition (FR) has become an important function in a wide-range of security and surveillance applications, involving computer networks, smartphones, tablets, IP cameras, etc. Capturing faces in still images or videos allows to perform non-intrusive authentication in applications where the user's co-operation is either impossible (video-surveillance in crowded environments) or to be limited (continuous authentication). For example, in the context of controlled access to critical information on computer network systems, the face modality may allow for a continuous and non-intrusive authentication [1]. After initial login, a FR system may enroll the authenticated user using facial images captured from the computer's built-in camera, and design a facial model.[1] The user's identity may then be periodically validated using facial images captured over time without requiring active co-operation (i.e. password prompt).

However, limited user co-operation as well as uncontrolled observation environments often make FR a challenging task. It is well known that the performance of state-of-the-art FR systems may be severely affected by changes in capture conditions (e.g. variations in illumination, pose and scale), as well as individual physiology [2, 3]. Moreover, such systems are usually initialized with a limited number of high-quality reference face captures, which may generate non-representative facial models (not modelling all possible variations) [4].

To account for such intra-class variations, several solutions have been investigated in the literature over the past decade. They can be organized into the following two categories:

1. Development of discriminative features that are robust to environmental changes [5, 6]. These techniques usually aim to develop facial descriptors insensitive to changes in capture conditions, to mitigate their effects on the recognition process.
2. Storage (or synthetic generation) of multiple reference images to cover the different capture conditions that could be encountered during operations [7, 8].

However, these approaches assume that FR is a stationary process, as they only rely on information available during enrolment sessions. In addition, depending on the

---

[1]Depending on the classification system, a facial model may be defined as either a set of one or more reference face captures (template matching) or a statistical model estimated from reference captures (statistical classification).

application, a single enrolment session is often considered as multiple ones are not always possible [9]. This prevents to integrate new concepts[2] that may emerge during operations as capture conditions and individuals physiology evolve over time (for example due to natural lighting conditions and ageing).

To address this limitation, adaptive biometric systems have been proposed in the literature [11], inspired by semi-supervised learning techniques for pattern recognition [12]. These systems are able to adapt facial models (sets of templates or classifier parameters) by exploiting (either on-line or off-line) faces captured during system operations. Common approaches in adaptive biometrics fall under *self-updating* and *co-updating*, depending on whether they rely on a single or multiple modalities. They usually either: 1) add novel captures to individual specific galleries [13], or 2), fuse new input data into common templates referred to as *super-templates*, containing all information [14, 15] for each modality (for example, virtual facial captures constructed with patches from operational data).

This chapter focuses on *self-updating* techniques with template matching systems for FR. These methods update template galleries using faces captured during operations that are considered highly confident, i.e. that produce very high matching scores (surpassing a self-updating threshold) [16]. Advantages and drawbacks of self-updating have been widely investigated [16, 17]. While these methods have been show to significantly improve the performance of biometric systems over time, an updating strategy only relying on matching score values may add redundant template to the galleries. This can significantly increase system complexity over time with information that do not necessarily improve performance, and also eventually reduce its response time during operations. To bound this complexity, template management methods (e.g. pruning) have been proposed in literature [16–18]. While clustering-based methods showed the most promising results, they remain computationally complex and thus not suited for seamless operations, if self-updating is performed frequently.

In this chapter, a survey of state-of-the-art techniques for adaptive FR using self-updating is presented, along with the key challenges facing these systems. An experimental protocol involving three real-life facial datasets (DIEE [19], FIA [20] and FRGC [21]) is proposed to evaluate the benefits and drawbacks of a self-updating methodology applied to a template matching system, with a particular focus on the management of system complexity. To address this challenge, a *context-sensitive* self-updating technique is proposed for template matching systems, combining a standard self-updating procedure and a change detection module. With this technique, only operational faces that were captured under different capture conditions are added to an individual's template gallery. More precisely, the addition of a new capture into the galleries depends on two conditions: (1) it's matching score is above the self-updating threshold (highly confident capture), and (2), the capture contains new information w.r.t. the samples already present in the gallery (i.e. captured under different conditions). This strategy allows to benefit from contextual information

---

[2] A *concept* can be defined as the underlying data distribution of the problem under specific operating conditions [10].

available in operational captures to limit the growth in system complexity. With this technique, one can avoid frequent uses of costly template management schemes, while still enhancing intra-class variation in facial models with relevant templates. A particular implementation of this proposed technique is considered for a basic template matching system, where changes are detected in illumination conditions.

The rest of this chapter is organized as follows. Section 2.2 provides a general survey of self-updating algorithms in the context of adaptive biometric systems. Then, Sect. 2.3 introduces the new *context-sensitive* self-updating technique based on the detection of changes in capture conditions, and Sect. 2.4 presents the proposed experimental methodology. Finally, experimental results are presented and discussed in Sect. 2.5.

2.2 Self-Updating for Face Recognition

2.2.1 A General Face Recognition System

Figure 2.1 presents a generic system for the recognition of faces in images (stills or video frames) captured from a camera. It is composed of four modules: segmentation, feature extraction, classification and decision. In addition, facial models of the N enrolled individuals are stored into the system, to be used by the classification module to produce matching scores for each individual.

During operations, faces are isolated in the image using the segmentation module, which produces the regions of interest (ROIs). Then, discriminant features are extracted from each ROI (e.g. eigenfaces [22] of local binary patterns [23]) to produce the corresponding pattern $\mathbf{d} = (d[1], \ldots, d[F])$ (with F the dimensionality of the feature space). This pattern is then compared to the facial model of each enrolled individual i by the classifier, which produces the corresponding matching scores $s_i(\mathbf{d})$, $(i = 1, \ldots, N)$.

The facial models are usually designed a priori using one or several reference patterns, from which the same features have been extracted, and their nature depends on the type of classifier used in the system. For example, with a template matcher, a facial model of an individual i can be a gallery of one or several reference patterns

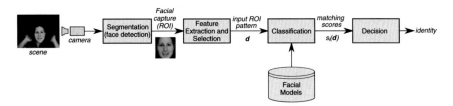

Fig. 2.1 General FR system trained for N individuals

$\mathbf{r}_{i,j}$ ($j = 1, \ldots, J$), in which case matching scores for each operational pattern $\mathbf{d}$ would be computed from distance measures to these patterns. Classification may also be performed using neural networks (e.g. multi-layer perceptrons [24] and ARTMAP neural networks [25]) or statistical classifiers (e.g. nave Bayes classification [26]), in which case the facial models would consist of parameters estimated during their training using the reference patterns $\mathbf{r}_{i,j}$ (e.g. neural networks weights, statistical distribution parameters, etc.).

Finally, the decision module produces a final response according to the application. For example, an identification system for surveillance may predict the identity of the observed individual with a maximum rule, selecting the enrolled individual with the highest matching score, while a verification system for access control usually confirms the claimed identity by comparing the corresponding matching score to a decision threshold.

2.2.2 Adaptive Biometrics

As mentioned earlier, the performance of FR systems can be severely affected by changes in capture conditions. Intra-class variations can be observed in the input data as a consequence of changes in capture conditions (scene illumination, facial pose angle w.r.t. the camera, etc.) or individuals physiology (facial hair, ageing, etc.). Such diversity is difficult to represent using the limited amount of reference captures used for initial facial model design. To address this limitation, adaptive biometric systems have been proposed in the literature, providing the option for continuous adaptation of the facial models using the operational data [9, 16].

Adaptation can be either supervised or unsupervised, depending on the labelling process of the operational data. In semi-supervised learning [27], the facial model of each individual enrolled to the system is updated using operational data labelled as the same individual by the classification system. For example, a gallery $\mathcal{G}_i$ of reference patterns may be augmented with highly confident operational input patterns $\mathbf{d}$ matched to the facial model of individual i. While this enables to refine facial models, the performance of such systems is strongly dependent on their initial classification performance. In addition, the integration of mislabelled captures could corrupt facial models, thus affecting the accuracy of the system for the corresponding individuals [16, 19].

An adaptive biometric system can also perform supervised adaptation, where the operating samples used to update the system are manually labelled, or obtained through some re-enrolment process [16]. While supervised adaptation may represent an ideal scenario with an error-free labelling process, human intervention is often costly or not feasible. Depending on the application, the ability to perform semi-supervised adaptation may be the only viable solution, which has lead to the development of various strategies to increase the robustness of such systems.

These techniques can be categorized as *self-update* [14, 15] and *co-update* techniques [16, 28] depending on whether a single or multiple modalities are considered for the update of facial models with highly confident patterns. This chapter focuses on *self-updating* methods for FR, where facial models are defined by galleries of reference patterns.

2.2.3 Self-Updating Methods

In the context of FR systems, self-updating methods update the facial models using only highly confident operational captures, i.e. with matching scores surpassing a very high threshold, to prevent possible corruptions due to misclassification.

2.2.3.1 General Presentation

To illustrate this principle, it is applied to a template matching system, presented in Fig. 2.2. In this system, inspired by [29], the facial model of each individual i is designed by storing initial reference patterns from a labelled dataset into a gallery $\mathcal{G}_i = \{\mathbf{r}_{i,1}, \mathbf{r}_{i,2}, \ldots\}$ (in this case, the terms *pattern* and *template* are used indiscriminately). To simplify the notation, the remainder of this section will omit the subscript i and only consider one individual, as this methodology can be extended to many with individual specific galleries and thresholds.

Algorithm 1 Self-update algorithm for adapting template galleries.

Inputs: - $\mathcal{G} = \{\mathbf{r}_1, \ldots, \mathbf{r}_J\}$ // *Gallery with initial templates*
       - $\mathcal{D} = \{\mathbf{d}_1, \ldots, \mathbf{d}_L\}$ // *Unlabelled adaptation set*

Outputs: - $\mathcal{G}' = \{\mathbf{r}_1, \ldots, \mathbf{r}_{J'}\}, J' \geq J$ // *Updated Gallery*

1: Estimate updating threshold $\gamma^u \geq \gamma^d$ from $\mathcal{G}$
2: $\mathcal{G} \leftarrow \mathcal{G}'$ // *Initialization with previous state*
3: **for** $l = 1, \ldots, L$ **do**   // *For all samples* $d_l \in \mathcal{D}$
4:     **for** $j = 1, \ldots, J$ **do**   // *For all references* $r_j \in \mathcal{G}$
5:       $s_j(\mathbf{d}_l) \leftarrow similarity\_measure(\mathbf{d}_l, \mathbf{r}_j)$ // *Prediction score for each reference*
6:     **end for**
7:     $S(\mathbf{d}_l) \leftarrow \max\limits_{j \in [1,J]} \{s_j(\mathbf{d}_l)\}$ // *Maximum fusion of scores*
8:     **if** $S(\mathbf{d}_l) \geq \gamma_d$ **then**
9:       Output positive prediction
10:      **if** $S(\mathbf{d}_l) \geq \gamma^u$ **then**
11:        $\mathcal{G}' \leftarrow \mathcal{G}' \cup \mathbf{d}_l$ // *Add the sample which similarity surpasses* γ^u *to the gallery*
12:      **end if**
13:    **end if**
14: **end for**

Algorithm 1 presents a generic algorithm for self-updating a template gallery $\mathcal{G}$ with several reference patterns $\mathbf{r}_j$ ($j = 1, \ldots, J$). During operations, the system is

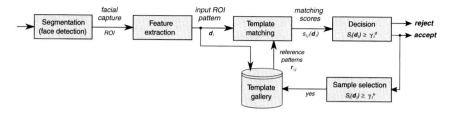

Fig. 2.2 A FR system based on template matching that allows for self-update

presented with an unlabelled data set $\mathscr{D}$ of L facial captures. For each sample $\mathbf{d}_l$, similarity measures to each reference $\mathbf{r}_j$ in the galley are used to compute the set of matching scores $s_j(\mathbf{d}_l)$ $(j = 1, \ldots, J)$. Then, the final score $S(\mathbf{d}_l)$ is computed as a combination of $s_j(\mathbf{d}_l)$ (e.g. the maximum fusion rule), and positive prediction is output if it surpasses the decision threshold γ^d. Finally, the sample selection module relies on a stricter updating threshold γ^u (usually $\gamma^u \geq \gamma^d$), updating the gallery $\mathscr{G}$ with $\mathbf{d}_l$ if $S(\mathbf{d}_l) \geq \gamma^u$, i.e. if the prediction has a high degree of confidence.

2.2.3.2 Challenges

While self-updating methods have been shown to improve system accuracy over time, the adaptation of facial models using operational data might be detrimental, and the selection of the updating threshold is critical [16]. To prevent a decline in classification performance, the use of a strict updating threshold may enable to reduce the probability of updating facial models with misclassified patterns [13, 15, 30]. However, it has been argued that updating with only highly confident patterns may result in the addition of redundant information in the galleries, and thus a marginal gain in performance at the expense of a considerable increase in system complexity [16].

In addition, operational samples with more drastic changes are less likely to generate classification scores surpassing the updating threshold, preventing the classification system to assimilate this new information. To address this limitation, *co-updating* methods have been proposed to benefit from complementary biometric systems [16, 28]. Each system is initialized with reference templates from a different source (or different features extracted from the same source), and performs classification of operational input data. In the same way as self-updating techniques, each system selects highly confident samples based on an updating threshold, but this information is also shared with other systems. If the classification score of one system surpasses its updating threshold, the others will also consider the corresponding samples as highly confident, and perform adaptation. This enables to increase the probability of updating with different but genuine operational data, by relying on the supposition that a drastic change on one source is not necessarily observed on others. A recent model has been proposed to estimate optimal amounts of samples and iterations to improve system's performance under specific updating constraints [31]. This model has shown to be effective under the stringent hypothesis of 0 % false alarm rate for the updating threshold of both systems. While *co-updating* is usually applied with multiple biometric traits, it could also be applied in, for example, a FR scenario involving multiple cameras. In this situation, relying on multiple points of view could mitigate the effect of disruptions such as motion blur that would be less likely to affect every camera at the same time.

Finally, system complexity is a critical issue for template matching systems in live FR. The ability to operate seamlessly depends on the computational complexity of the recognition operation, which is usually directly related to gallery sizes. Several template management strategies have been proposed to limit complexity in self-updating systems. In [18], template replacement strategies have been experimented to perform self-update in a constrained environment, where the maximum number of templates in a gallery is fixed by the user. When the maximum size is reached, several criteria have been experimented to determine which obsolete template can be replaced, such as FIFO, LFU and clustering algorithms. Among them, the clustering algorithm MDIST showed the most promising results, reducing the number of impostors samples by maintaining a gallery with very close samples. While these methods enable to compromise between system performance and complexity, they remain computationally costly, and may interfere with seamless long-term operations. Once the maximum gallery size is reached, such process would have to be performed for each new highly confident template, thus increasing system response time. To reduce these occurrences, operational data containing redundant information should be filtered out during operations. This would limit the self-updating process to only operational templates with relevant information, i.e. templates improving intra-class variability in facial models.

2.3 Self-Updating Driven by Capture Conditions

This chapter introduces a new self-updating method that efficiently self-updates facial models based on capture conditions. This methodology is illustrated using a template matching system performing self-updating, as presented in [29]. As discussed in the previous sections, such methodology can significantly improve the overall classification performance through a better modelling of intra-class variations , specifically in applications exhibiting significant variations in capture conditions (e.g. continuous authentication using webcams). However, updating the galleries with only highly confident inputs may not always provide new and beneficial information, as those samples are usually well-classified by the system, which could lead to an unnecessary increase in system complexity (e.g. the number of reference patterns stored in the galleries) [16]. While this complexity can be mitigated with template management techniques [18], frequent gallery filtering may interfere with seamless operations over time.

To address this limitation, this section proposes a *context-sensitive* self-updating technique that integrates a template filtering process during operations. It is designed to ensure that only highly confident data captured under novel conditions are added to template galleries, thus limiting the growth in memory complexity with redundant samples. In fact, in FR, intra-class variations in facial appearance are often related to changes in capture conditions (e.g. environmental illumination, facial pose, etc.) [2, 3], and such information can be detected during operations. Following this intuition, when a highly confident ROI pattern surpasses the updating threshold, non-discriminative information related to capture conditions are extracted to evaluate whether it has been captured under different conditions that of the reference templates already stored in the gallery. If not, the pattern is discarded, and the gallery is not augmented.

2.3.1 Framework for Context-Sensitive Self-Update

The diagram of a general template matching system that employs the new *context-sensitive* technique is presented in Fig. 2.3. It augments the system presented in Fig. 2.2 with an additional decision module to detect changes in capture conditions.

In the same way than standard self-updating systems, when presented with a unlabelled data set $\mathcal{D} = \{\mathbf{d}_1, \ldots, \mathbf{d}_L\}$, this system first selects highly confident samples to perform adaptation of the template gallery $\mathcal{G}_i$, i.e. the set $\mathcal{D}' = \{\mathbf{d}_{l'} | S_i(\mathbf{d}_{l'}) \geq \gamma_i^u\}$. Then, an additional test is performed on these samples, only to select a final subset captured under novel capture conditions. To extract additional non-discriminative information, the individual galleries are augmented with the input ROIs $\mathbf{R}_{i,j}$ from which the reference patterns $\mathbf{r}_{i,j}$ are extracted. The augmented galleries are stored as $\mathcal{G}_i = \{\{\mathbf{R}_{i,1}, \mathbf{r}_{i,1}\}, \{\mathbf{R}_{i,2}, \mathbf{r}_{i,2}\}, \ldots\}$. This additional measurement enables to maximize the intra-class variation of the galleries, while mitigating their growth by rejecting redundant information. For example, contextual information such as environmental illumination or facial pose w.r.t. the camera can be measured on ROIs, to be compared with ROIs in the galleries.

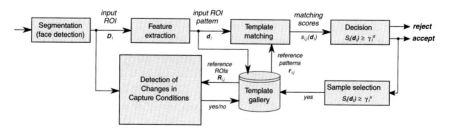

Fig. 2.3 A template matching system that integrates context-sensitive self-updating

2.3.2 A Specific Implementation

As a basic example of the framework presented in Fig. 2.3, a particular implementation is proposed. It relies on the detection of changes in illumination conditions.

2.3.2.1 A Template Matching System

For classification, a standard template matching system is considered. For each individual i, a dedicated facial model is stored as a template gallery $\mathscr{G}_i = \{\{\mathbf{R}_{i,1}, \mathbf{r}_{i,1}\}, \{\mathbf{R}_{i,2}, \mathbf{r}_{i,2}\}, \ldots, \{\mathbf{R}_{i,J_i}, \mathbf{r}_{i,J_i}\}\}$, as well as user-specific decision γ_i^d and updating γ_i^u thresholds.

For each input ROI isolated through segmentation, the corresponding pattern $\mathbf{d}_l$ is extracted using a Multi-Bloc Local Binary Pattern (LBP) [23] algorithm. Features for block sizes of 3×3, 5×5 and 9×9 pixels are computed and concatenated with the grayscale pixel intensity values, and PCA is used to reduce the dimensionality to $F = 32.^3$ The matching score for each individual i is then computed as follows:

$$S_i(\mathbf{d}_l) = \frac{1}{J_i} \cdot \sum_{j=1}^{J_i} s_{i,j}(\mathbf{d}_l) = \frac{1}{J_i} \sum_{j=1}^{J_i} \frac{\left[\sqrt{F} - d_{Eucl}(\mathbf{d}_l, \mathbf{r}_{i,j})\right]}{\sqrt{F}} \tag{2.1}$$

where $d_{Eucl}(\mathbf{d}_l, \mathbf{r}_{i,j})$ is the Euclidean distance between input pattern $\mathbf{d}_l$ and template $\mathbf{r}_{i,j}$ (with $j = 1, \ldots, J_i$) and J_i the total number of templates in $\mathscr{G}_i$. The matching scores $s_{i,j}(\mathbf{d}_l)$ are here computed as the normalized opposite to the distance $d_{Eucl}(\mathbf{d}_l, \mathbf{r}_{i,j})$ (a score of 1 is achieved for a null distance). The final matching score $S_i(\mathbf{d}_l)$ is obtained from the combination of these scores using the average fusion rule.

Finally, the system outputs a positive prediction for individual i if $S_i(\mathbf{d}_l) \geq \gamma_i^d$, and selects $\mathbf{d}_l$ as a highly confident face capture for individual i if $S_i(\mathbf{d}_l) \geq \gamma_i^u$.

2.3.2.2 Detecting Changes in Capture Conditions

In Fig. 2.3, for each individual i, the input ROIs $\mathbf{D}_l$ corresponding to highly confident operational captures are compared to the reference ROIs $\mathbf{R}_{i,j}$ ($j = 1, \ldots, J_i$) stored in the galleries, and asses whether the capture conditions are novel enough to justify an increase in complexity. The universal image quality index Q [32] is considered to measure the distortion between $\mathbf{D}_l$ and each reference ROI $\mathbf{R}_{i,j}$. This measure is a particular case of the structural similarity index measure (SSIM) presented in [33]. It can be written as a product of the three factors—loss of correlation, luminance distortion and contrast distortion:

---

^{3}This value has been determined experimentally as an optimal trade-off between accuracy and computational complexity using a nearest neighbour classifier with Euclidean distance.

$$Q(\mathbf{R}_{i,j}, \mathbf{D}_l) = \frac{\sigma_{\mathbf{R}_{i,j}, \mathbf{D}_l}}{\sigma_{\mathbf{R}_{i,j}} \cdot \sigma_{\mathbf{D}_l}} \cdot \frac{2\bar{\mathbf{R}}_{i,j} \cdot \bar{\mathbf{D}}_l}{\bar{\mathbf{R}}_{i,j}^2 + \bar{\mathbf{D}}_l^2} \cdot \frac{2\sigma_{\mathbf{R}_{i,j}} \cdot \sigma_{\mathbf{D}_l}}{\sigma_{\mathbf{R}_{i,j}}^2 + \sigma_{\mathbf{D}_l}^2} \tag{2.2}$$

where $\bar{\mathbf{R}}_{i,j}$ and $\bar{\mathbf{D}}_l$ are the average images, $\sigma_{\mathbf{R}_{i,j}}$ and $\sigma_{\mathbf{D}_l}$ their variances, and $\sigma_{\mathbf{R}_{i,j}, \mathbf{D}_l}$ the covariance.

To accommodate spatial variations in image distortion, statistical features for Eq. 2.2 may be measured locally. A local quality index $Q(\mathbf{R}_{i,j}[k], \mathbf{D}_l[k])$ is thereby calculated, where $\mathbf{D}_l[k]$ ($\mathbf{R}_{i,j}[k]$) corresponds to a window of $\mathbf{D}_l$ ($\mathbf{R}_{i,j}$) sliding from the top-left corner to the bottom-right corner for a total of K steps. These local measurements can then be combined into the global quality index GQ following:

$$GQ(\mathbf{R}_{i,j}, \mathbf{D}_l) = \frac{1}{K} \sum_{k=1}^{K} Q(\mathbf{R}_{i,j}[k], \mathbf{D}_l[k]) \tag{2.3}$$

In this chapter, the proposed template filtering strategy is implemented through a detection of changes in ROI illumination conditions only. For that intent, the second term of the quality index Q (see Eq. 2.2) is considered, to compute the global luminance quality (GLQ) following:

$$GLQ(\mathbf{D}_l, \mathbf{R}_{i,j}) = \frac{1}{K} \sum_{k=1}^{K} LQ(\mathbf{R}_{i,j}[k], \mathbf{D}_l[k]) = \frac{1}{K} \sum_{k=1}^{K} \frac{2 \cdot \bar{\mathbf{D}}_l[k] \cdot \bar{\mathbf{R}}_{i,j}[k]}{\bar{\mathbf{D}}_l[k]^2 + \bar{\mathbf{R}}_{i,j}[k]^2} \tag{2.4}$$

where the local luminance quality measurements LQ measure the proximity of the average luminance between each window. Highly confident captures $\mathbf{D}_l$ are then used to update the gallery $\mathcal{G}_i$ if and only if

$$\frac{1}{J_i} \sum_{j=1}^{J_i} GLQ(\mathbf{D}_l, \mathbf{R}_{i,j}) \geq \gamma_i^c \tag{2.5}$$

with γ_i^c the capture condition threshold, computed as the average GLQ between all the references captures in $\mathcal{G}_i$.

2.4 Simulation Methodology

This section presents several experimental scenarios involving three real-world FR databases. The proposed simulations emulate realistic FR applications of different orders of complexity, with variations in capture conditions. The objective is to observe and compare the performance of new and reference self-updating techniques under different operation conditions, and within a basic template matching system described in Sect. 2.3.2.

Table 2.1 Summary of the three experimental scenarios

Dataset	Scenario	# enrolled individuals	# enrolment sessions	# ROIs per batch	Sources of variation
DIEE	Continuous authentication	49	6	10	Illumination, expression
FIA	Video-surveillance	10	3	69	Illumination, expression, pose, resolution, ageing, scaling, blur
FRGC	Wide-range identification	187	16	22	Illumination, expression, ageing

2.4.1 Face Recognition Databases

Three publicly available FR databases are considered for simulation. To standardize the experimental protocol, each database is separated into six different batches for all individuals. These scenarios are summarized at the end of Sect. 2.4.1, in Table 2.1.

2.4.1.1 Multi-Modal Dipartimento di Ingegneria Elettrica Ed Elettronica

The multi-modal Dipartimento di Ingegneria Elettrica ed Elettronica[4] (DIEE) dataset [19] regroups face and fingerprint captures of 49 individuals. In this study, only facial captures are considered. For each individual, 60 facial captures have been acquired over 6 sessions at least 3 weeks apart, with 10 captures per session. The collection process spaned over a period of 1.5 years.

For simulations, the facial captures or each individuals are separated into six batches corresponding to the capture sessions. ROIs have been extracted with a semi-manual process [34]: an operator first selected the eyes in each frame, and the cropped region was then determined as the square of size $2d * 2d$ (d being the distance between the eyes), with the eyes located at the position $(d/2, d/4)$ and $(3 \cdot d/2, d/4)$. In this process, faces have been rotated to align the eyes to minimize intra-class variations [35], and then normalized to a size of 70×70 pixels.

This dataset was explicitly collected to evaluate the performance of self-update and co-training algorithms. Over the 6 sessions, gradual changes can be observed in facial pose, orientation and illumination (see examples in Fig. 2.4). While these changes generate visible differences in facial captures, the position of the individuals and their distance to the camera are controlled. For this reason, this dataset represents the

---

[4]Department of Electrical and Electronic Engineering.

Fig. 2.4 DIEE dataset. An example of randomly chosen facial captures for two individuals

easiest problem in this study, simulating an application of continuous authentication of individuals over a computer network.

2.4.1.2 CMU Faces in Action

The Carnegie Mellon University Faces In Action (FIA) dataset [20] contains a set of 20 s videos for 221 participants, mimicking a passport checking scenario in both indoor and outdoor environments. Videos have been captured in three separate sessions of 20 s at least 1 month apart, with 6 Dragonfly Sony ICX424 cameras (640×480 pixel resolution, 30 images per second). Cameras were positioned at 0.83 m of the subjects, mounted on carts at three different horizontal angles ($0°$ and $\pm 72.6°$), with two focal lengths (4 and 8 mm) each.

In this chapter, only ROIs captured during the indoor sessions, and using the frontal camera with 8 mm focal length are considered. ROIs have been extracted using the OpenCV implementation of Viola-Jones face and eye detection algorithm [36]. In the same way than with DIEE, faces have been rotated to align the eyes [35], and normalized to a size of 70×70 pixels. For simulations, sequences from each session have been divided into two sub-sequences, in order to organize the facial captures into six batches.

This dataset simulates an open-set surveillance scenario as found in face re-identification applications. A restrained subset of 10 individuals of interest are monitored, but in an environment where a majority of ROIs are capture from non-target individuals. The 10 individuals of interest enrolled to the systems have been chosen with two experimental constraints: (1) the individuals must be present in all capture sessions and (2) at least 30 ROIs per session have been extracted by the face detection algorithm.

Faces in this data set have been captured in semi-controlled capture conditions, where the individuals entered the scene and walked to stop at the same distance from the cameras, and talked while moving their head with natural movements until the end of the session. In addition to variations in illumination and facial expressions, ROIs also incorporate variations in pose, resolution (scaling), motion blur and ageing (see Fig. 2.5).

Fig. 2.5 FIA dataset. An example of randomly chosen facial captures for two individuals

2.4.1.3 Face Recognition Grand Challenge

The Face Recognition Grand Challenge (FRGC) dataset as been collected at University Notre Dame [21]. In this chapter, the still face images of this dataset are considered. They were captured over an average of 16 sessions for 222 individuals for the training subset, and up to 22 sessions for the validation one, using a 4 Megapixels Canon camera. Each session contains four controlled and two uncontrolled captures, with significantly different illumination and expression.

Overall, 187 individuals have been selected for experiments, for which more than 100 ROIs are available (around 133 in average). In the same way than with the other datasets, six batches of the the sane size have been created for each individual, respecting the temporal relation between the capture sessions. ROIs have been extracted in the same way than with the DIEE dataset [34], using the position of the eyes already available in the FRGC dataset.

This dataset simulates a wide-range identification application, with multiple re-enrolment sessions where a very limited amount of reference templates are captured. Recurring and unpredictable changes in illumination and facial expression emerge in the operational environment in every capture session (see Fig. 2.6).

Fig. 2.6 FRGC dataset. An example of randomly chosen facial captures for two individuals

2.4.2 Protocol

The following three template matching systems are experimentally compared in this chapter:

1. **baseline system**, performing template matching in the same way as in Fig. 2.3, but without any adaptation of the template galleries $\mathscr{G}_i$. User-specific decision thresholds γ_i^d are stored for decision.
2. standard **self-updating** system, updating the template galleries $\mathscr{G}_i$ with highly confident ROI patterns, which scores surpass user-specific updating thresholds γ_i^u, and decision thresholds γ_i^d.
3. proposed **context-sensitive self-updating** system, only updating the template galleries $\mathscr{G}_i$ with highly confident samples that also passed the concept change test (Eq. 2.5), using user-specific updating γ_i^u, capture condition γ_i^c and decision thresholds γ_i^d.

2.4.2.1 Simulation Scenario

The scenario described below is considered for each database. At each time step $t = 1, \ldots, 6$, and for each individual $i = 1, \ldots, N$, the performance of the baseline and the two self-updating systems updated with batch $b_i[t-1]$ is evaluated on batch $b_i[t]$. The self-updating systems are updated, and then tested with batch $b_i[t+1]$, and so on. A pseudocode of the simulation process is presented in Alg. 2.

Algorithm 2 Protocol for simulations.

1: **for** $i = 1, \ldots, N$ **do**   *// Initialization of the galleries $\mathscr{G}_i$ for each individual*
2:   $\mathscr{G}_i \leftarrow$ first 2 patterns of $b_i[1]$
3: **end for**
4: **for** $i = 1, \ldots, N$ **do**   *// Initialization of thresholds for each individual*
5:   Evaluate update and decision thresholds γ_i^u and γ_i^d using negative distribution estimation
6:   Initialize change detection threshold γ_i^c as the average GLQ measure between each ROI in $\mathscr{G}_i$
7: **end for**
8: **for** $i = 1, \ldots, N$ **do**   *// Processing of remaining samples from $b_i[1]$*
9:   Estimate genuine scores using remaining samples from $b_i[1]$
10:   Estimate impostor samples using a random selection of impostor samples
11:   Update gallery
12:   Update thresholds
13: **end for**
14: **for** $t = 2, \ldots, 6$ **do**   *// Remaining data blocks*
15:   **for** $i = 1, \ldots, N$ **do**   *// Each individual*
16:     Estimate genuine scores using remaining samples from $b_i[t]$
17:     Estimate impostor samples using a random selection of impostor samples
18:     Update gallery
19:     Update thresholds
20:   **end for**
21: **end for**

For each system, the individual galleries $\mathcal{G}_i$ are initialized with the two first samples of the corresponding initial batches $b_i[1]$. For *context-sensitive* self-updating, corresponding ROIs are also stored to compute GLQ measures during operations (see Eq. 2.4). Then, the initial values of the decision thresholds γ_i^d are computed using negative distribution estimation; each gallery $\mathcal{G}_i$ is compared to every other gallery to generate negative scores, and a threshold γ_i^d is chosen as the highest possible value respecting an operational false alarm constraint. For the self-updating variants, the updating threshold γ_i^u is initialized in the same way, and for the *context-sensitive* self-updating system, γ_i^c is computed as the average GLQ measure between each ROI in $\mathcal{G}_i$.

Then, for each system, performance is evaluated using the remaining patterns from $b_i[t]$ to compute genuine scores, and a random selection of impostor patterns for the impostor scores. For the DIEE and FRGC datasets, impostor patterns for each individual are randomly selected among batches from other individuals. In the case of the FIA dataset, impostor patterns are selected from the non-target dataset individuals during the same session. To avoid any bias in performance evaluation, the same amount of impostor and genuine patterns are considered.

Finally, using genuine and impostor patterns, the self-updating systems galleries are updated according to their updating strategies, and the thresholds are re-estimated using the same methodology. This scenario is then reproduced for the remaining five batches.

2.4.2.2 Performance Measures

For each system, performance is measured with average true positive rate (tpr) and false positive rate (fpr) for each individual. These are, respectively, the proportion of genuine patterns correctly classified over the total number of genuine patterns (tpr), and the proportion of impostor patterns classified as genuine over the total number of negative patterns (fpr). These measures depend on the decision thresholds γ_i^d, computed during update to respect a given fpr constraint.

System complexity is also presented, as the average number of templates in the galleries. In addition, facial model corruption due to the addition of misclassified templates in the galleries is presented as the ratio of impostor over genuine templates. Following Doddingtons classification [37], only the 10 galleries with the highest ratio are presented, to focus on lamb-type individuals which are easy to imitate.

Finally, a constraint of $fpr = 5\%$ has been chosen to compute the decision thresholds γ_i^d. In addition, for each scenario, the updating thresholds γ_i^u correspond to an ideal $fpr = 0\%$ and a laxer $fpr = 1\%$. For each performance measure, results are presented as the average and standard deviation values for every enrolled individual, computed using a Student distribution and a confidence interval of 10%.

2.5 Simulation Results

2.5.1 Continuous User Authentication with DIEE Data

Figure 2.7 presents the average performance results of the baseline, self-updating and *context-sensitive* self-updating techniques within the template matching system described in Sect. 2.3.2. Results are presented for the ideal $fpr = 0\%$ updating thresholds for the self-updating techniques.

While all three systems present similar fpr (between 7 and 17%) in Fig. 2.7a, a significant differentiation can be observed in the tpr with batches five and six (Fig. 2.7b). In fact, the introduction of batch five generates a decline in tpr performance for the baseline system (from $43.5 \pm 5.7\%$ down to $33.0 \pm 6.5\%$), that ends at $tpr = 39.3 \pm 6.6\%$ at batch six. On the other hand, the self-updating and *context-sensitive* self-updating systems exhibit a moderate decline (respectively, from $47.5 \pm 6.1\%$ to $41.3 \pm 6.7\%$), and end at a higher performance of $tpr = 46.3 \pm 7.3\%$.

Even with a $fpr = 0\%$ updating threshold, it can be observed that this FR scenario benefits from a self-updating strategy, as the addition of up to an average 13.7 ± 2.4 templates in the galleries (see Fig. 2.7c) enabled to increase the system's performance. In addition, despite the limited amount of captures (10 per session), the filtering of the *context-sensitive* self-updating system enabled to maintain a comparable level of

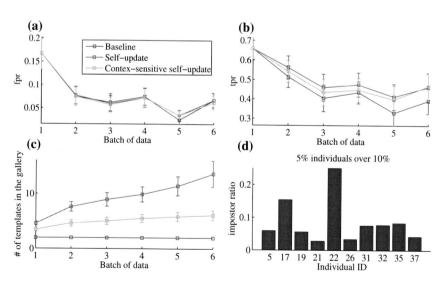

Fig. 2.7 Simulation results with DIEE dataset where the updating threshold is selected for $fpr = 0\%$ **a** false positive rate, **b** true positive rate, **c** system complexity, **d** impostor ratio in the galleries of the *top* 10 lambs-like individuals

performance with a significantly lower amount of templates in the gallery, ending at an average of 6.1 ± 0.9 templates.

Despite the relative simplicity of this scenario and the restrictive updating threshold, impostor templates have been incorrectly added to the galleries during the updating process. Following Doddington's analysis, the ratio of impostor over genuine templates in the galleries of the top 10 lamb individuals (i.e. the individuals with the highest ratio) is presented in Fig. 2.7d. While 95% of the galleries contain under 10 % of impostor samples, two lamb-like individuals (ID 17 and 22) stand out with over 10 and 20 % impostor samples in their galleries.

Figure 2.8 presents the average performance results for the *fpr* $= 1\%$ updating thresholds for the self-updating techniques. An overall performance increase is shown for the self-updating methods. A higher tpr is observed throughout the entire simulation, ending at *tpr* $= 55.0 \pm 7.7\%$ for self-updating, and *tpr* $= 50.8 \pm 7.0\pm$ for *context-sensitive* self-updating (see Fig. 2.8b).

While results with self-updating are higher in this application, it is important to note that improvements come at the expense of a doubled average gallery size (see Fig. 2.8c), as well as an increase in the impostor ratio (see Fig. 2.8d), 20 % of the galleries are composed by more than 10 % impostor templates). Comparing these ratios with the the previous ones (in Fig. 2.7), it is apparent that this increase is not connected to specific lamb-type individuals, but to all the enrolled individuals. This underlines the importance of updating thresholds, specifically for long-term operations where the impostor ratio would be likely to grow exponentially as the facial models become corrupted.

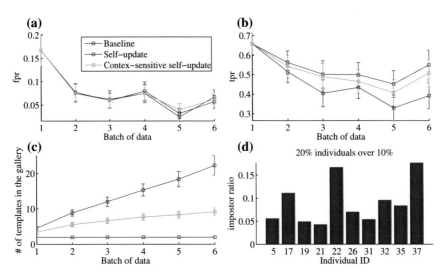

Fig. 2.8 Simulation results with DIEE dataset where the updating threshold is selected for *fpr* $= 1\%$ **a** false positive rate, **b** true positive rate, **c** system complexity, **d** impostor ratio in the galleries of the *top* 10 lambs-like individuals

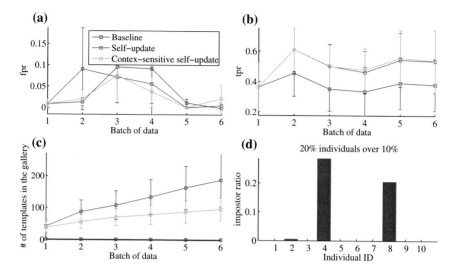

Fig. 2.9 Simulation results with FIA dataset where the updating threshold is selected for *fpr* = 0 %
a false positive rate, **b** true positive rate, **c** system complexity, **d** impostor ratio in the galleries of
the *top* 10 lambs-like individuals

2.5.2 Video-Surveillance with FIA Data

Figure 2.9 presents the average performance results for the *fpr* = 0 % updating
thresholds for the self-updating techniques. In this scenario, involving more sources
of variations in capture conditions than the DIEE dataset (see Table 2.1), the benefits
of a self-updating strategy are more significant, as the self-updating systems exhibit
a significantly higher tpr during the entire simulation (see Fig. 2.9b). From batch
two to six, the self-updating systems are stable close to *tpr* = 60 % (both ending
at 53 ± 20 %), while the baseline system remains close to *tpr* = 40 % (ending at
38.1 ± 17.2 %).

As a consequence of the more complex nature of a semi-controlled surveillance
environment as well as the higher number of facial captures, performance improvements come at the expense of significantly larger galleries than with the DIEE dataset
(see Fig. 2.9c), ending at an average of 188 ± 83 templates for self-update, and
97 ± 35 templates for *context-sensitive* self-update. It can still be noted that the
filtering strategy of the *context-sensitive* self-update technique enables to maintain a
comparable level of performance, for gallery sizes approximately two times smaller.

Among 10 individuals of interest, 2 lamb-like individuals (ID 4 and 8) can be
identified, with an impostor ratio over 20 % (see Fig. 2.9d). Despite the added complexity of a semi-constrained environment, the higher number of faces captured in
video streams enables a better definition of facial models of target individuals during the first batch. This explains that impostor templates have only be added to two
difficult lamb-type individuals, and not all the galleries.

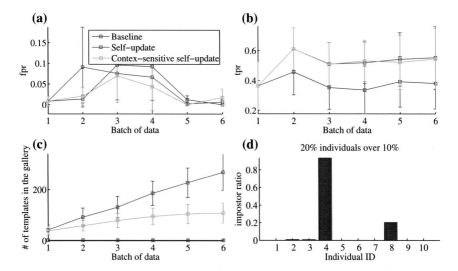

Fig. 2.10 Simulation results with FIA dataset where the updating threshold is selected for $fpr = 1\%$ **a** false positive rate, **b** true positive rate, **c** system complexity, **d** impostor ratio in the galleries of the *top* 10 lambs-like individuals

In Fig. 2.10b, it can be observed that a more relaxed $fpr = 1\%$ constraint for the updating threshold did not have a significant impact on the performance of self-updating systems. However, the average gallery size of the self-updating technique increased to end at 268 ± 71 templates, while the *context-sensitive* self-updating technique enabled to remain at a lower size of 109 ± 38 templates (see Fig. 2.10c), comparable to the $fpr = 0\%$ threshold results (see Fig. 2.9c). This observation reveals that a majority of the new templates added with the $fpr = 1\%$ thresholds contained redundant information, that was already present in the galleries. This underscores the benefits of the *context-sensitive* self-updating technique when operating with videos, where higher quantities of templates may be selected for self-updating. By reducing the number of updates, this technique enables to mitigate the growth in computational complexity of the prediction process as well as the need to use a costly template management system, without impacting system performance.

Impostor ratios in Fig. 2.10d show a significant increase for individual ID 8, which ends at 80%. This confirms the rapid addition of impostor templates to the galleries in long-term operations. In this video-surveillance scenario, where more facial captures are presented to the system (compared to the DIEEE scenario), the gallery of lamb-like individual four is updated with a larger amount of impostor templates at the beginning of the simulation. This gallery then keeps attracting impostor templates over time, which reduces the pertinence of the facial model.

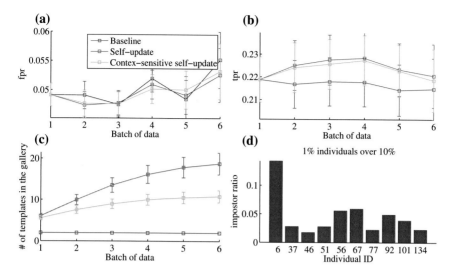

Fig. 2.11 Simulation results with FRGC dataset where the updating threshold is selected for $fpr = 0\%$ **a** false positive rate, **b** true positive rate, **c** system complexity, **d** impostor ratio in the galleries of the top 10 lambs-like individuals

2.5.3 Unconstrained Face Recognition with FRGC Data

Figure 2.11 presents the average performance results for the $fpr = 0\%$ updating thresholds for the self-updating techniques. It can be observed in Fig. 2.11b that this scenario represents a significantly harder FR problem, as all three systems perform below $tpr = 23\%$ during the entire simulation. In addition, despite the increase in average gallery size up to, respectively, 18.8 ± 2.7 and 10.8 ± 1.5 templates for the self-update and *context-sensitive* self-update techniques (see Fig. 2.11c), only a marginal performance gain can be observed. The two self-updating systems end at $tpr = 22.1 \pm 1.4\%$ and $tpr = 21.9 \pm 1.4\%$, while the baseline case exhibits a $tpr = 21.5 \pm 1.4\%$.

A bigger impact can be observed in Fig. 2.12b, presenting tpr performance of the three systems for the $fpr = 1\%$ updating threshold. From batch two to six, the two self-updating cases present significantly higher tpr performance, both ending at $tpr = 25.6 \pm 1.5\%$. However, as in the previous scenarios, this performance gain comes at the expense of a significantly higher system complexity. Both systems with self-update end with, respectively, 82.4 ± 5.2 and 41.6 ± 2.0 templates in the galleries (see Fig. 2.12c). The average impostor ratio also increased significantly, as 18% of the galleries contain more than 10% impostor templates (see Fig. 2.12d), while only 1% of the galleries were in this situation with the $fpr = 0\%$ updating threshold.

Results are related to the nature of the scenario presented in Sect. 2.4.1.3. The multiple enrolment sessions (up to 16), where small numbers of ROI were captured

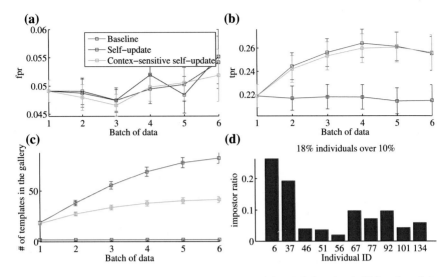

Fig. 2.12 Simulation results with FRGC dataset where the updating threshold is selected for *fpr* = 1 % **a** false positive rate, **b** true positive rate, **c** system complexity, **d** impostor ratio in the galleries of the top 10 lambs-like individuals

(6 ROIs), favour the presence of genuine captures that are different enough to fail the updating threshold test. Fewer than 20 templates per individuals have been added to self-updating galleries with *far* = 0 % self-updating threshold (see Fig. 2.11c), despite the presence of more than 100 genuine samples in batches. In addition, the systems are initialized with the first capture session, where only four controlled stills are available to build the facial model before processing uncontrolled captures in future sessions. This prevents the generation of representative facial models, that either reject a majority of genuine templates, or accept a significant amount of impostor templates depending on the updating threshold (see Fig. 2.12d).

Despite the improved performance achieved using self-updating techniques, this dataset raises the limitations of using a self-updating system relying on a two-threshold update strategy in complex environments, with limited reference data and uncontrolled variations in capture conditions.

2.5.4 Summary and Discussions

In all experimental results, the following general observations have emerged:

1. Both self-updating techniques generate a significant and stable performance boost over time.
2. The template filtering strategy of the proposed *context-sensitive* self-updating technique significantly reduces system complexity. The galleries are approxi-

mately two times smaller than a standard self-updating system, without impacting performance.

3. Using a less stringent constraint of $fpr = 1\%$ for the updating threshold does not always have an impact on the performance boost, but always increases system complexity as well as the number of impostor templates in the galleries.

While these observations remain valid for each scenario, a more precise analysis reveals potential limitations of these approaches depending on the represented application.

In a semi-controlled FR application with limited changes mainly caused by illumination and expression (DIEE dataset), benefits of a self-updating techniques are quite clear. In fact, despite the increase in the number of impostor samples in the gallery, a significant performance boost can be observed when a more relaxed updating threshold is selected.

In the case of a video-surveillance scenario involving a higher amount of impostor individuals not modelled by the system (FIA dataset), a more relaxed updating threshold did not show any performance improvement, despite a doubled average gallery size for the self-updating technique (while the *context-sensitive* self-update technique prevented any increase in average gallery size). While the overall performance was not lowered, the gallery of one specific individual was severely affected, ending with around 80% of impostor samples. In such scenario, involving multiple causes of variation (face angle, resolution, motion blur, etc.) as well as a greater amount of impostor individuals, manual intervention may be necessary at regular intervals, to ensure that the gallery of some specific individuals (lambs) are not getting corrupted over time.

Finally, in the more complex scenario represented by the FRGC dataset, the performance gain observed with the self-updating techniques was considerably lower, even with the less stringent updating threshold of $far = 1\%$. In this scenario, systems are presented with significantly different samples in early operations (after the fourth image), as opposed to the DIEE and FIA scenarios (with, respectively, 10 and around 30 samples for a first session). In such application, a manual intervention may be required at the early stages of operations, to ensure that the facial models are initialized with enough representative templates to be able to keep updating over time.

2.6 Conclusion

Despite the advances in feature extraction and classification techniques, face recognition in changing environments remains a challenging pattern recognition problem. Changes in capture condition or individuals physiology can have a significant impact on a system performance, where initial facial models are often designed with a limited amount of reference templates, and frequent re-enrolment sessions are not always possible. Adaptive classification techniques have been proposed in the past decade to address this challenge, relying on operational data to adapt the system over time. Among them, self-updating techniques have been proposed for automatic adaptation using highly confident captures labelled by the system. While this enables to automatically benefit from a considerable source of new information without requiring a costly manual updating process, these systems are particularity sensitive to their internal parameters. A trade-off between assimilation of new information and protection against the corruption of facial models with impostor templates has to be considered, as well as a limitation of system complexity over time. While template management techniques can be used to limit system complexity, they remain costly and may interfere with seamless operations.

In this chapter, self-updating methods have been surveyed in the context of a face recognition application with template matching. A *context-sensitive* self-update technique has been presented to limit the growth in system complexity over time, by relying on additional information related to the capture conditions. With this technique, only highly confident faces captured under new conditions are selected to update individual facial models, effectively filtering out redundant information. A specific implementation of a template matching system with *context-sensitive* self-update has been proposed, where changes are detected in illumination conditions. Proof-of-concept experimental simulations using thee publicly available face databases showed that this technique enables to maintain the same level of performance than a regular self-updating template matching system, with a significant gain in terms of memory complexity. Using additional information available in the face captures during operations, this technique allows to reduce the size of template galleries by half, effectively mitigating the computational complexity of the recognition process over time. In applications where memory footprint has to be restricted, this strategy would also limit the need to use costly template management techniques during operations.

However, application-specific limitations have been observed during simulations. When faced with recognition environments with significant variations, and a limited pool of reference patterns for initial enrolment, self-updating systems can be very sensitive to the initialization of their template galleries, as well as the updating threshold. A stricter updating rule may be required to prevent updating with impostor samples, which can significantly reduce the benefits of a self-updating strategy that would never detect any highly confident samples. In addition, while the proposed *context-sensitive* self-updating techniques enabled to significantly reduce system complexity, it relies on the storage of input ROIs in addition to reference patterns in the galleries, as well as an additional measurement during operations.

While self-updating techniques can significantly improve the recognition performance of face recognition systems, their implementation should always be tailored to the specificities of the application as well as the recognition environment. While human intervention can be reduced with automatic strategies, it will still play a critical role in certain applications, especially when dealing with significant variations in capture conditions. In those cases, occasional manual confirmation should be considered, in order to maintain the system's performance by adapting to abrupt changes.

References

1. Niinuma, N., Park, U., Jain, A.: Soft biometric traits for continuous user authentication. IEEE Trans. Inf. Forensics Secur. **5**(4), 771780 (2010)
2. Pagano, C., Granger, E., Sabourin, R., Marcialis, G., Roli, F.: Adaptive ensembles for face recognition in changing video surveillance environments. Inf. Sci. **286**, 75–101 (2014)
3. De-la Torre, M., Granger, E., Radtke, P.V., Sabourin, R., Gorodnichy, D.O.: Partially-supervised learning from facial trajectories for face recognition in video surveillance. Information Fusion (2014)
4. Pagano, C., Granger, E., Sabourin, R., Gorodnichy, D.O.: Detector ensembles for face recognition in video surveillance. In: Neural Networks (IJCNN), The 2012 International Joint Conference on, pp. 1–8. IEEE (2012)
5. De Marsico, M., Nappi, M., Riccio, D., Wechsler, H.: Robust face recognition for uncontrolled pose and illumination changes. IEEE Trans. Syst Man Cybern. **43**(1), 149–163 (2012)
6. Wright, J., Yang, A., Ganesh, A., Sastry, S.: Robust face recognition via sparse representation. IEEE Trans. Pattern Anal. Mach. Intell. **31**(2), 210–227 (2009)
7. Jafri, R., Arabnia, H.: A survey of face recognition techniques. J. Inf. Process. Syst. **5**(2), 41–68 (2009)
8. Li, S.Z., Jain, A.K.: Handbook of Face Recognition, 2nd edn. Springer Publishing Company, Incorporated (2011)
9. Rattani, A.: Adaptive biometric system based on template update procedures. Ph.D. thesis. University of Cagliari, Italy (2010)
10. Narasimhamurthy, A., Kuncheva, L.I.: A framework for generating data to simulate changing environments. In: Proceedings of the 25th IASTED International Multi-Conference: Artificial Intelligence And Applications, pp. 84–389 (2007)
11. Roli, F., Didaci, L., Marcialis, G.: Adaptive biometric systems that can improve with use. In: Ratha, N., Govindaraju, V. (eds.) Advances in Biometrics, pp. 447–471. Springer, London (2008)
12. Nagy, G.: Classifiers that improve with use. Procs. Conference on Pattern Recognition and Multimedia, vol. 103, pp. 79–86. IEICE, Tokyo, Japan (2004)
13. Roli, F., Marcialis, G.L.: Semi-supervised pca-based face recognition using self-training. In: Structural, Syntactic, and Statistical Pattern Recognition, pp. 560–568. Springer (2006)
14. Jiang, X., Ser, W.: Online fingerprint template improvement. Pattern Anal. Mach. Intell. IEEE Trans. **24**(8), 1121–1126 (2002)
15. Ryu, C., Kim, H., Jain, A.K.: Template adaptation based fingerprint verification. In: 18th IEEE International Conference on Pattern Recognition, ICPR 2006, vol. 4, pp. 582–585 (2006)
16. Rattani, A., Freni, B., Marcialis, G.L., Roli, F.: Template update methods in adaptive biometric systems: A critical review. Lecture Notes in Computer Science (included Lecture Notes in Artificial Intelligence and Lecture Notes in Bioinformatics) **5558**, 847–856 (2009)
17. Marcialis, G.L., Rattani, A., Roli, F.: Biometric template update: an experimental investigation on the relationship between update errors and performance degradation in face verification. In: Structural, Syntactic, and Statistical Pattern Recognition, pp. 684–693. Springer (2008)
18. Freni, B., Marcialis, G.L., Roli, F.: Replacement algorithms for fingerprint template update. In: Image Analysis and Recognition, pp. 884–893. Springer (2008)
19. Rattani, A., Marcialis, G.L., Roli, F.: A multi-modal dataset, protocol and tools for adaptive biometric systems: a benchmarking study. Int. J. Biometrics **5**(4), 266–287 (2013)
20. Goh, R., Liu, L., Liu, X., Chen, T.: The cmu face in action (fia) database. In: Proceedings of Analysis and Modelling of Faces and Gestures. Second International Workshop. AMFG 2005, pp. 255–63. Springer, Berlin (2005)
21. Phillips, P.J., Flynn, P.J., Scruggs, T., Bowyer, K.W., Chang, J., Hoffman, K., Marques, J., Min, J., Worek, W.: Overview of the face recognition grand challenge. In: Proceedings of the 2005 Conference on Computer Vision and Pattern Recognition (CVPR'05), pp. 947–954. IEEE (2005)

22. Turk, M.A., Pentland, A.P.: Face recognition using eigenfaces. In: Proceedings of 1991 IEEE Computer Society Conference on Computer Vision and Pattern Recognition, pp. 586–91. IEEE (1991)
23. Ahonen, T., Hadid, A., Pietikainen, M.: Face description with local binary patterns: application to face recognition. IEEE Trans. Pattern Anal. Mach. Intell. **28**(12), 2037–2041 (2006)
24. Riedmiller, M.: Advanced supervised learning in multi-layer perceptrons from backpropagation to adaptive learning algorithms. Comput. Stan. Interfaces **16**(3), 265–278 (1994)
25. Carpenter, G.A., Grossberg, S., Reynolds, J.H.: Artmap. Supervised real-time learning and classification of nonstationary data by a self-organizing neural network. Neural Netw. **4**(5), 565–588 (1991)
26. Duda, R.O., Hart, P.E.: Pattern recognition and scene analysis (1973)
27. Zhu, X.: Semi-supervised learning literature survey (2005)
28. Rattani, A., Marcialis, G.L., Roli, F.: Capturing large intra-class variations of biometric data by template co-updating. In: Computer Vision and Pattern Recognition Workshops, 2008. CVPRW'08. IEEE Computer Society Conference on, pp. 1–6. IEEE (2008)
29. Rattani, A., Marcialis, G.L., Roli, F.: Self adaptive systems: An experimental analysis of the performance over time. In: Computational Intelligence in Biometrics and Identity Management (CIBIM), 2011 IEEE Workshop on, pp. 36–43. IEEE (2011)
30. Liu, X., Chen, T., Thornton, S.M.: Eigenspace updating for non-stationary process and its application to face recognition. Pattern Recogn. **36**(9), 1945–1959 (2003)
31. Didaci, L., Marcialis, G.L., Roli, F.: Analysis of unsupervised template update in biometric recognition systems. Pattern Recogn. Lett. **37**, 151–160 (2014)
32. Wang, Z., Bovik, A.C.: A universal image quality index. Signal Process. Lett. IEEE **9**(3), 81–84 (2002)
33. Wang, Z., Bovik, A.C., Sheikh, H.R., Simoncelli, E.P.: Image quality assessment: from error visibility to structural similarity. IEEE Trans. Image Process. **13**(4), 600–612 (2004)
34. Marcialis, G.L., Roli, F., Fadda, G.: A novel method for head pose estimation based on the "virtuvian man". Int. J. Mach. Learn. Cybern. **5**(11), 111–124 (2014)
35. Gorodnichy, D.O.: Video-based framework for face recognition in video. In: Proceedings of 2nd Canadian Conference on Computer and Robot Vision, pp. 330–338 (2005)
36. Viola, P., Jones, M.J.: Robust real-time face detection. Int. J. Comput. Vis. **57**, 137–154 (2004)
37. Doddington, G., Liggett, W., Martin, A., Przybocki, M., Reynolds, D.: Sheep, goats, lambs and wolves: A statistical analysis of speaker performance in the nist 1998 speaker recognition evaluation. Tech. rep, DTIC Document (1998)

Chapter 3
Handling Session Mismatch
by Semi-supervised-Based Co-training
Scheme

Norman Poh, Joseph Kittler and Ajita Rattani

Abstract Co-training-based semi-supervised learning scheme has been shown to be
a viable training strategy for handling the mismatch between training and test samples. For co-training-based multimodal biometric systems, classical semi-supervised
learning strategies such as self-training and co-training may not have fully exploited
the advantage of a multimodal fusion, notably due to the fusion module. For this
reason, this chapter discusses a novel semi-supervised training strategy known as
fusion-based co-training that generalizes the classical co-training such that it can use
a trainable fusion classifier. Experiments on the BANCA face and speech database
show that this proposed strategy is a viable approach. In addition, we also resolve
the issue of how to select the decision threshold for adaptation. In particular, we find
that a strong classifier, including a multimodal system, may benefit better from a
more relaxed threshold, whereas a weak classifier may benefit better from a more
stringent threshold.

3.1 Introduction

Biometric person authentication remains a challenging problem for two key reasons.
First, there are very few enrolment samples to train the model for a particular user.
Second, there is often significant variation between the samples used for enrolment
and those used for authenticating the user (which are the test or query samples). This
problem is sometimes referred to as a train–test mismatch. This mismatch between the
enrolment and test samples, or session mismatch, is caused by a number of factors.

N. Poh (✉) · J. Kittler
University of Surrey, Surrey, UK
e-mail: normanpoh@ieee.org

J. Kittler
e-mail: j.kittler@surrey.ac.uk

A. Rattani
Michigan State University, East Lansing, MI, USA
e-mail: ajita@msu.edu

© Springer International Publishing Switzerland 2015   35
A. Rattani et al. (eds.), *Adaptive Biometric Systems*,
Advances in Computer Vision and Pattern Recognition,
DOI 10.1007/978-3-319-24865-3_3

The data acquisition process is vulnerable to these variations. For instance, face images can easily be effected by changes in illumination while speech signals can be corrupted by environmental noise such as passing cars or other people speaking [16]. Other factors include the nature of biometric traits being biological samples that can alter temporarily or permanently, for instance, as a result of ageing, diseases, or treatment to a disease.

An important consequence of the above factors is that a biometric reference—which could be a template or a statistical model—cannot be expected to fully and automatically cope with all possible sources of variation. A promising learning paradigm to solve the above problem is *semi-supervised learning*. In this paradigm, a biometric system is initialized with correctly labelled samples and then (as a classifier) attempts to label the test samples and considers these samples as potential training samples; the initialization is the only part that is supervised, hence, the name semi-supervised. If the samples are labelled correctly, the system can indeed capture the variation of the test conditions. On the other hand, if an impostor's sample is labelled as being genuine, the resultant system may perform significantly worse over a period of time. While there exists a large body of literature on semi-supervised learning [30] and concept drift dealing with general adaptive pattern recognition systems [22, 26, 28], the biometric problem deserves a dedicated treatment of its own, considering that a biometric system is potentially rolled out in a very large scale and may persist throughout the life time of a person.

Two well-known semi-supervised learning strategies are *self-training* and *co-training*. In *self-training*, a unimodal biometric system (face or speech) attempts to update its parameters using the highly confidently classified test samples [13]. In *co-training*, the mutual and complementary help of two classifiers is utilized to adapt the parameters (references) to the intra-class variation of the input operational data. Commonly, highly confidently classified sample for one of the classifiers is used to adapt the parameters of both the classifiers [16, 17, 19, 30]. However, this process does not involve a fusion module, i.e., a module that combines two or more outputs of the constituent biometric systems. Since a fusion module is often employed in multimodal biometrics, it is natural to consider a *fusion-based co-training* scheme. This is the main subject of study in this paper. In addition, in order to understand the mechanics and behaviour of the strategies, we also investigated a cross-training strategy where the inferred labels from one modality are added as training samples for another modality. The four different schemes, namely, self-training, co-training, cross-training (as a control), and the proposed fusion-based co-training, are shown in Fig. 3.1.

We validate our experiments using the publicly available bimodal face and speech BANCA database [2]. This database contains three acquisition conditions, namely controlled, adverse and degraded conditions. With reference to the controlled conditions, the adverse ones are due to acquisition in a noisy environment, whereas the degraded ones are due to the use of a different acquisition device. The impact of these three conditions is clearly visible in Fig. 3.2.

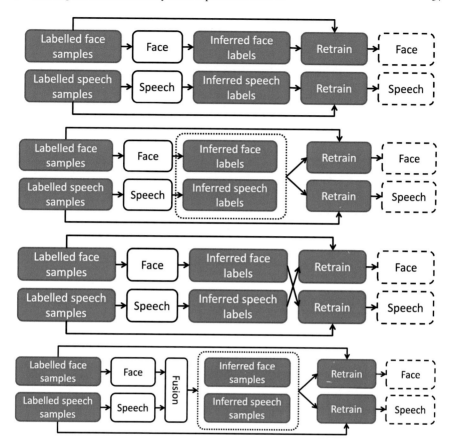

Fig. 3.1 A data flow diagram of the extended view of semi-supervised learning strategies. Boxes drawn with *solid lines* denote initial models, whereas those drawn with *long-dashed lines* denote updated models. Boxes drawn with a *short-dashed lines* imply a union operator

Our contributions can be summarized as follows: First, we propose a novel co-training-based fusion algorithm. Second, we investigate the unresolved problem of threshold determination for adaptive systems. Third, we experimentally validate the proposed adaptation strategy with self-training, co-training, and a control system known as "cross-training".

This chapter is organized as follows. Section 3.1.1 presents the methodology; Sect. 3.2, database and experiments; Sect. 3.3, results; and Sect. 3.4, conclusions. This section presents the four schemes shown in Fig. 3.1

Fig. 3.2 The three scenarios of the BANCA database

Algorithm 1 The self-training algorithm

- Given: labelled data $\mathcal{L}$ and unlabelled data $\mathcal{U}$
- Loop until stop criterion satisfied:

  - Train g1 using $\mathcal{L}$
  - Label $\mathcal{U}$ using g1 to obtain $\mathcal{U}_*$
  - Add the highly confident self-labelled samples from $\mathcal{U}_*$ to $\mathcal{L}$
  - Remove the self-labelled examples from $\mathcal{U}$

3.1.1 Methodology

3.1.1.1 Revisiting Self-training and Co-training

In self-training, a biometric system, say g1, attempts to infer labels from an unlabelled dataset $\mathcal{U}$. If the labels are inferred with sufficiently high confidence, they are incorporated into a labelled dataset, $\mathcal{L}$. The labelling process is repeated until no more labels can be inferred by this way or stopped at a predetermined number of iterations [3, 22]. Algorithm 1 describes this procedure more formally.

In the original co-training algorithm [3], two biometric systems, say g1 and g2, attempt to infer labels from $\mathcal{U}$ independently. The confidently labelled samples are then added to $\mathcal{L}$. This procedure is described in Algorithm 2. The main difference between Algorithms 1 and 2 is that the information between the two modality experts are not shared in self-training, whereas in co-training, the experts work collaboratively.

3.1.1.2 Fusion-Based Co-training

We observe that in the original co-training algorithm, the *union* of two inferred labels by two algorithms is used simultaneously. Hence, this operation can be interpreted as an OR fusion rule.

Rather than using the OR fusion rule, in multimodal biometrics, it is common to use a trainable classifier such as logistic regression. In essence, the OR rule is an example of decision-level fusion, whereas logistic regression is an example of

---

Algorithm 2 The original co-training algorithm

- Given: labelled data $\mathcal{L}$ and unlabelled data $\mathcal{U}$
- Loop until stop criterion satisfied:

  - Train g1 using $\mathcal{L}$
  - Train g2 using $\mathcal{L}$
  - Label $\mathcal{U}$ using g1 to obtain $\mathcal{U}_*^1$
  - Label $\mathcal{U}$ using g2 to obtain $\mathcal{U}_*^2$
  - Add the highly confident self-labelled examples in $\mathcal{U}_*^1 \bigcup \mathcal{U}_*^2$ to $\mathcal{L}$
  - Remove the self-labelled examples from $\mathcal{U}$

Algorithm 3 The fusion-based co-training algorithm

- Given: labelled data $\mathcal{L}$ and unlabelled data $\mathcal{U}$
- Loop until stop criterion satisfied:

  - Train g1 using $\mathcal{L}$
  - Train g2 using $\mathcal{L}$
  - Train f using $\mathcal{L}$ (with a twofold cross-validation; see text)
  - Label $\mathcal{U}$ using f to obtain $\mathcal{U}_*$
  - Add the highly confident self-labelled examples in $\mathcal{U}_*$ to $\mathcal{L}$
  - Remove the self-labelled examples from $\mathcal{U}$

---

score-level fusion. Score-level fusion is generally better than decision-level fusion because the former considers the *confidence* of expert output, as reflected by the *absolute value* of the matching score. This piece of information is simply not taken into account in decision-level fusion.

Because of the above nature, combining a strong expert with a weak expert at the decision level is not always beneficial [19]. In comparison, the score-level fusion can still optimally utilize the strength of both experts even with this unbalanced performance, in the Bayes sense [26].

The fusion-based co-training algorithm is shown in Algorithm 3. The main difference in this algorithm compared to self-training and co-training is that the fusion-based co-training requires an additional step in order to train the fusion classifier. This step is not needed if the fusion classifier used is based on fixed rules such as sum and product.

Since a trainable fusion classifier is used here (which is logistic regression), one has to ensure that the data samples (scores) used to train the fusion classifier should not be the same as those used to train the baseline experts, otherwise the resultant trained fusion classifier will be overly optimistically biased (i.e., the expert outputs will appear more confident than they should be).

To avoid the positive bias, one can adopt k-fold cross-validation. For instance, let k be two. One can then partition $\mathcal{L}$ into two non-overlapping sets, say $\mathcal{L}_1$ and $\mathcal{L}_2$. Then, one trains a base expert on $\mathcal{L}_1$ and generates scores using $\mathcal{L}_2$. Similarly, one trains another base expert on $\mathcal{L}_2$ and obtains scores using $\mathcal{L}_1$. The union of the

---

Algorithm 4 The cross-training algorithm

- Given: labelled data $\mathcal{L}$ and unlabelled data $\mathcal{U}$
- Let $\mathcal{L}^1=\mathcal{L}$ and $\mathcal{L}^2=\mathcal{L}$
- Loop until stop criterion satisfied:

  - Train g1 using $\mathcal{L}^2$
  - Label $\mathcal{U}$ using g1 to obtain $\mathcal{U}_*^2$
  - Train g2 using $\mathcal{L}^1$
  - Label $\mathcal{U}$ using g2 to obtain $\mathcal{U}_*^1$
  - Add the highly confidently labelled samples from $\mathcal{U}_*^1$ to $\mathcal{L}^1$ and $\mathcal{U}_*^2$ to $\mathcal{L}^2$
  - Remove the examples just labelled (i.e., $\mathcal{U}_*^1$ and $\mathcal{U}_*^2$) from $\mathcal{U}$

---

resultant scores obtained from $\mathcal{L}_1$ and $\mathcal{L}_2$ are used to train the fusion classifier. In this way, the training scores are unbiased from the fusion's perspective.

3.1.1.3 Cross-training

Since the difference between self-training and co-training is principally due to whether or not information is exchanged, it is instructive to study *how* this information is exchanged. A possible intermediate way of information exchange is to retrain one classifier using the labels inferred by another classifier. This gives rise to the *cross-training* algorithm, as shown in Algorithm 4.

3.2 Database and Experiments

3.2.1 Database

We shall use the BANCA database [2], because it is a multimodal biometric database and for both modalities, simulated authentication sessions were recorded under controlled, adversed, and degraded conditions, as shown in Fig. 3.2. The clear distinction in the sample quality is ideal for conducting controlled experiments in our case. This database contains videos of 52 subjects reading text-prompted sentences as well as answering short questions. The subjects are divided into two groups of 26 subjects, which is designed for a twofold cross-validation experiment.

A consequence of this BANCA database setting is that the face verification problem becomes extremely challenging, compared to the speaker verification problem. This is because in both the adverse and degraded conditions, the noise due to the environmental conditions affecting the speech modality, which consists of indoor recordings, is still relatively unimportant in comparison with the face modality.

A novel aspect concerning the usage of this database, unlike precedent efforts in [11] or [10], is that *video sequences* are actually used here, rather than *still images* extracted from the video sequence.

3.2.2 BANCA Protocols

The BANCA database contains 12 sessions, recorded under 3 conditions. Sessions 1–4 were recorded under the controlled conditions; 5–9, degraded conditions; and 11–12, adverse conditions. The BANCA database comes with its own experimental protocols known as "P" (for partitioned) and "G" (for general), which are applicable to experiments with training (also known as the enrolment set in biometrics) and test datasets. However, in experiments involving adaptation, it is important to designate another set of data for adaptation. For this reason, we have modified P and G protocols slightly, leading to an adaptive protocol as shown in Table 3.1. The first line in Table 3.1 shows the number of examples *for each client* in each partition of data. The second line shows the exact session numbers used to constitute the respective partition of data as well as the conditions under which the data (video) sample is obtained (controlled, adverse or degraded).

In order to calculate the number of samples for each partition of the data, one multiplies the numbers in rows "# match samples" and "# non-match samples" of Table 3.1 by 26, because there are 26 enrollment subjects in each group of subjects, recalling that the 52 subjects have been divided into two groups with balanced gender composition.

The enrollment data partition normally does not contain non-match samples. However, the numbers indicated here (with †) are the number of samples used for training the background model or for feature extraction (e.g., principal component analysis); they come with the BANCA English database.

The experiments are designed to compare five adaptive settings: (1) no adaptation, which serves as a baseline; (2) self-training; (3) cross-training; (4) fusion-based co-training; and, finally (5) supervised adaptation, which establishes the upper bound of the achievable performance. The training set of the non-adaptive system consists of only the enrollment partition of the data; the adaptive partition of data is not used at all. On the other hand, for the supervised training, the data consists of the enrollment and adaptive partitions of the data, i.e., the combined sessions {1, 5, 9}, are used to train the system. This corresponds exactly to the original BANCA protocol. In all five settings, the test partition is reserved uniquely for evaluating the system performance. Each of these settings is evaluated on the unimodal face, unimodal speech, and bimodal fusion systems.

When performing verification on the BANCA database, errors were measured in *Half Total Error Rate (HTER)* $HTER = \frac{FRR+FAR}{2}$, where FRR is the ratio of true clients that were falsely rejected and FAR is the ratio of impostors that were falsely accepted.

Table 3.1 Adaptive protocol for the BANCA database

Datasets	Enrolment			Adaptation			Test		
conditions	c	a	d	c	a	d	c	a	d
# match samples	1	0	0	0	1	1	3	3	3
Session	{1}				{5}	{9}	{2, 3, 4}	{6, 7, 8}	{10, 11, 12}
# non-match samples	20 †	20 †	20 †	0	25	25	4	4	4
Session	{1}	{5}	{9}		{5}	{9}	{2, 3, 4}	{6, 7, 8}	{10, 11, 12}

Note c = controlled, a = adverse, d = degraded. Each row shows the number of match or non-match samples *for each client*. There are 26 clients in each group, and there are two groups according to the Banca protocols. †: the numbers indicated here are the *background* models

3.2.3 Baseline Systems

The face and speaker verification baseline systems (also referred to as experts) are Bayesian classifiers whose class-conditional densities are approximated using Gaussian Mixture Models (GMMs) with the maximum *a posteriori* adaptation [24]. This is a long-standing state-of-the-art classifier for the speaker verification, but since then, it has also been successfully used for the face verification problem [4]. The face verification problem can benefit from this approach, mainly thanks to parts-based local feature descriptors, which represent an face image by a set of overlapping or non-overlapping blocks of image. For each block of image, its texture is described using a *local feature descriptor.*

Let $\mathbf{X} \equiv \{\mathbf{x}_i | i = 1, \ldots, N\}$ be a sequence of N feature frames and each feature frame be denoted by $\mathbf{x}_i$ (for the i-th frame). For the face modality, a feature frame is a vector containing the DCT coefficients of a block of image. For the speech modality, a feature frame contains mel-scale cepstral coefficients. These features are a short-term representation of spectral envelopes filtered by a set of filters motivated by the human auditory system.

Let $p(\mathbf{x}|\omega_o)$ be the likelihood function of the world or background model and $p(\mathbf{x}|\omega_j)$ be the model for the claimed identity $j \in \{1, \ldots, J\}$. In parts-based face or speaker verification, both $p(\mathbf{x}|\omega_o)$ and $p(\mathbf{x}|\omega_j)$, for any j, are estimated using a Gaussian Mixture Model (GMM) [24]. The world model is first obtained from a large pool of sequences $\{\mathbf{X}\}$ contributed by a large and possibly separate population of users (possibly from an external database than the one used for enrollment/testing). Each client-specific model is then obtained by adapting the world model upon presentation of the enrollment data of a specific user/client.

If the score y is greater than a pre-specified threshold, one declares that the query data $\mathbf{X}$ belongs to the model j. Hence, this will result in an acceptance decision. Otherwise, one rejects the hypothesis and hence rejects the identity claim.

The speaker verification classifier used here differs from the face one in the following ways. First, the variability across sessions are removed, thanks to a standard technique called factor analysis [24]. This technique is applied to all training and test data prior to building a (client-specific) GMM model.

3.2.4 Threshold Determination

Before showing the result, there is another crucial aspect: threshold determination. One simple strategy is to use all possible thresholds. For this purpose, we choose three levels of threshold: a "relaxed", a "moderate" and a "stringent" threshold for adaptation.

A principled way of determining the three levels of threshold *by their confidence* is to map the threshold onto a probabilistic scale, e.g., the probability being a client given the expert output (say y):

$$\text{confidence } (y) = P(C|y)$$

After this mapping process, also known as *score calibration*, we simply compare confidence (y) with the adaptation threshold, Δ_{adapt}. If confidence (y) exceeds the threshold, then the corresponding sample is added to the labelled set $\mathcal{L}$. Thus, the three levels of threshold can be taken as $\{0.25, 0.50, 0.75\}$, respectively, for the relaxed, moderate and stringent thresholds, respectively.

Throughout our experiments, the posterior probability is estimated via logistic regression, which is trained on the held out group (i.e., when testing the BANCA G1 group of users, the data of G2 is used for training). The logistic regression is expressed by

$$P(C|y) = \frac{1}{1 + \exp(-(w_1 y + w_0))}$$

where y is the output of the unimodal system (face or speech). For the bimodal fusion, we use

$$P(C|y) = \frac{1}{1 + \exp(-(w_2 y_2 + w_1 y_1 + w_0))}$$

instead, where y_i is expert output i and w_i is its associated weight. The weight parameters are estimated using the expectation maximization principle. The realized algorithm is known as "gradient-ascent" [24].

3.3 Results

The experiments are divided into two parts: unimodal (face and speech) systems and bimodal fusion. For each case, we shall plot only the half total error rate (HTER) and pooled DET curves (over G1 and G2 groups of subjects). For the assessment here, the threshold used here minimizes the equal error rate on the development dataset.

3.3.1 Unimodal Systems

We will explain the different face systems tested, as it is clear that the speech systems can be explained in exactly the same way.

- **Face baseline**: this is the original non-adaptive system; it effectively assumes that the adaptive partition of data simply does not exist.
- **Face self-train**: this is a self-training system that attempts to infer labels from the adaptive partition of data. The inferred data samples are used to augment the original enrollment partition of data for training.
- **Face co-trained by speech**: this is a cross-training setting where a face system is trained by the labels obtained from the speech system.

- **Face co-trained by fusion**: This system is trained by the labels obtained from the fusion system–logistic regression in our case.
- **Face supervised**: this is a supervised system, where the labels of the adaptive partition of data are known; this represents the lowest achievable error rates on this dataset.

The speech systems are obtained in a similar way. For example, "speech system co-trained by face" refers to the speech system that is trained by samples in the adaptation set whose labels are inferred by the face system.

The results of each of the face and speech systems are shown in Fig. 3.3 using DET curves and their corresponding HTER points are shown in Fig. 3.4. Each shows the DET curves of the five systems, along with three levels of calibrated threshold (by its confidence), for the face and speech modality separately. We observe the following:

- The self-training speech system benefits from the relaxed threshold (0.25).
- On the other hand, the self-training face system benefits from the more stringent threshold (0.75).
- The speech system that is co-trained by face degrades significantly in performance, compared to the baseline non-adaptive system.
- On the other hand, the face system benefits from co-training by the speech system.
- The fusion-based co-training performs most optimally with the relaxed threshold (0.25).

These observations are all consistent in supporting the case that fusion-based co-training is better than cross-training (face cross-trained by speech or speech cross-trained by face).

It is interesting to observe that the self-training strategy degrades the face system but improves the speech system. This suggests that an already strong (good) system is likely to benefit from self-training, whereas a weak system may further degrade in performance with self-training due to the inclusion of wrongly labelled samples.

3.3.2 Multimodal System

Figure 3.5 presents the DET curves of the different multimodal systems. The figure again reveals a similar trend, except that the difference of performance between the supervised system and the best performing co-training system (at 0.25 adaptation threshold) is significantly larger here.

Fig. 3.3 DET curves of (**a**) face and (**b**) speech adaptive experiments. In the legend, "c.b." stands for co-trained or cross-trained. "F-LR" stands for Fusion using Logistic Regression (LR). Therefore, "Face c.b. Speech" reads "Face cross-trained by Speech" and "Face c.b. F-LR" reads "Face co-trained by Fusion using Logistic Regression". The HTER of these DET curves is summarized in Fig. 3.4. (The figure is best viewed in colour)

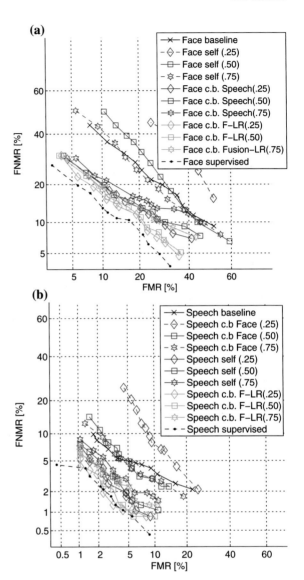

(a)

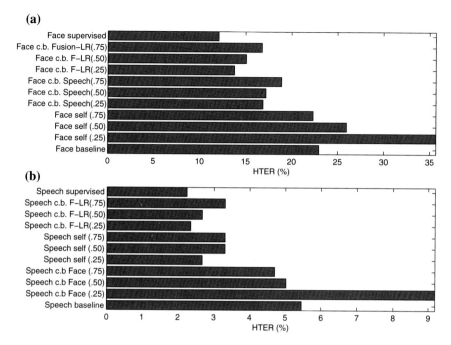

(b)

Fig. 3.4 HTER of (**a**) face and (**b**) speech systems. Note the scale difference in the X-axis, indicating that the speech system is several times better than the face system for this dataset

Fig. 3.5 Pooled DET curves the 11 fusion systems. (The figure is best viewed in colour)

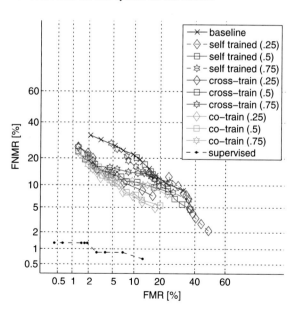

3.4 Conclusions

Semi-supervised learning is a general methodology that can be applied to any classification problem. For multimodal biometric systems, however, the classical co-training strategy would not have fully exploited the capability of the fusion system. For this reason, it is worthy to investigate fusion-based co-training as an addition to the existing adaptive strategy. Recognizing that the classical co-training corresponds to the OR rule fusion, we investigate a fusion-based co-training using a trainable fusion scheme that is implemented using logistic regression. We compare this method with two other strategies, namely (1) "cross-training" in order to understand the impact of cross modality training without fusion, and (2) training with complete supervision in order to assess the most optimistic performance. The experiments reveal that determining the effective threshold for adaptation remains a challenging problem. For a weak system (with relatively low classification accuracy), using a more stringent threshold appears to be better. On the other hand, for a strong classifier (with relatively high accuracy), using a relaxed adaptation threshold appears to benefit the system.

References

1. Akhtar, Z., Ahmed, A., Erdem, C.E., Foresti, G.L.: Biometric template update under facial aging. In: Proceedings of IEEE Symposium on Computational Intelligence in Biometrics and Identity Management (IEEE SSCI-CIBIM) (2014)
2. Bailly-Bailli'ere, E., Bengio, S., Bimbot, F., Hamouz, M., Kittler, J., Marithoz, J., Matas, J., Messer, K., Popovici, V., Poree, F., Ruiz, B., Thiran, J.-P.: The BANCA database and evaluation protocol. In: International Conference Audio- and Video-Based Biometric Person Authentication (AVBPA) (2003)
3. Blum, A., Mitchell, T.: Combining labeled and unlabeled data with co-training. In: Proceedings of the Eleventh Annual Conference on Computational Learning Theory, University of Wisconsin-Madison, pp. 92–100, New York, NY, USA (1998)
4. Cardinaux, F., Sanderson, C., Marcel, S.: Comparison of MLP and GMM classifiers for face verification on XM2VTS. In: Proceedings of International Conference on Audio- and Video-based Biometric Person Authentication, pp. 1058–1059 (2003)
5. Flynn, P.J., Bowyer, K.W., Phillips, P.J.: Assessment of time dependency in face recognition: An initial study. In: Proceedings of 4th International Conference on Audio and Video based Biometric Person Authentication, pp. 44–51 (2003)
6. Franco, A., Maio, D., Maltoni, D.: Incremental template updating for face recognition in home environments. Pattern Recogn. **43**, 2891–2903 (2010)
7. Jiang, X., Ser, W.: Online fingerprint template improvement. IEEE Tran. PAMI **8**, 1121–1126 (2002)
8. Lanitis, A., Taylor, C.J., Cootes, T.F.: Toward automatic simulation of aging effects on face images. IEEE Tran. Pattern Anal. Mach. Intell. **24**(4), 442–455 (2002)
9. Liu, X., Chen, T., Thornton, S.M.: Eigenspace updating for non-stationary process and its application to face recognition. Pattern Recognition, pp. 1945–1959 (2003)
10. Messer et al.: Face authentication competition on the banca database. In: International Conference on Biometric Authentication, pp. 815 (2004)
11. Messer et al.: Face authentication test on the banca database. Int. Conf. Pattern Recogn. **4**, 523–532 (2004)

12. Pavani, S.K., Sukno, F.M., Butakoff, C., Planes, X., Frangi, A.F.: A confidence based update rule for self-updating human face recognition systems. In: Proceedings of International Conference on Biometrics (ICB), pp. 151–160 (2009)
13. Poh, N., Wong, R., Kittler, J., Roli, F.: Challenges and research directions for adaptive biometric recognition systems. In: Proceedings of International Conference on Biometrics (ICB), pp. 753–764 (2009)
14. Poh, N., Kittler, J., Marcel, S., Matrouf, D., Bonastre, J.F.: Model and score adaptation for biometric systems: coping with device interoperability and changing acquisition conditions. In: Proceedings of 20th International Conference on Pattern Recognition (ICPR), pp. 1229–1232 (2010)
15. Poh, N., Kittler, J., Rattani, A., Tistarelli, M.: Group-specific score normalization for biometric systems. In: Proceedings of IEEE Conference on Computer Vision and Pattern Recognition Workshops (CVPRW), pp. 38–45 (2010)
16. Poh, N., Rattani, A., Roli, F.: Critical analysis of adaptive biometric systems. IET Biometrics 1(4), 179–187 (2012)
17. Rattani, A.: Adaptive biometric system based on template update procedures, Ph.D. thesis. University of Cagliari, Italy (2010)
18. Rattani, A., Marcialis, G.L., Roli, F.: Biometric template update using the graph mincut: a case study in face verification. In: Proceedings 6th IEEE Biometric Symposium (2008)
19. Rattani, A., Freni, B., Marcialis, G.L., Roli, F.: Template update methods in adaptive biometric systems: a critical review. In: Proceedings of International Conference on Biometrics (ICB), pp. 847–857 (2009)
20. Rattani, A., Marcialis, G.L., Roli, F.: An experimental analysis of the relationship between biometric template update and the doddingtons zoo in face verification. In: Proceedings of 14th International Conference on Image Analysis and Processing (ICIAP'09), pp. 434–442, Vietri sul Mare (Italy) (2009)
21. Rattani, A., Marcialis, G.L., Roli, F.: Temporal analysis of biometric template update procedures in uncontrolled environment. In: Proceedings of 16th International Conference on Image Analysis and Processing (ICIAP'11), pp. 595–604, Raveena, Italy (2011)
22. Rattani, A., Marcialis, G.L., Roli, F.: Biometric system adaptation by self-update and graph-based techniques. J. Vis. Lang. Comput 24(1), 1–9 (2013)
23. Rattani, A., Marcialis, G.L., Granger, E., Roli, F.: A dual-staged classification-selection approach for automated update of biometric templates. In: International Conference on Pattern Recognition (ICPR), pp. 2972–2975 (2014)
24. Reynolds, D.A., Quatieri, T., Dunn, R.: Speaker verification using adapted gaussian mixture models. Digital Signal Process. 10(1–3), 1941 (2000)
25. Roli, F., Marcialis, G.L.: Semi-supervised pca-based face recognition using self training. In: Proceedings of Joint ICPR Int'l workshop on S+SSPR (2006)
26. Roli, F., Didaci, L., Marcialis, G.L.: Template co-update in multimodal biometric systems. In: International Conference on Biometrics (ICB), pp. 1194–1202 (2007)
27. Ryu, C., Hakil, K., Jain, A.: Template adaptation based fingerprint verification. In: Proceedings of International Conference on Pattern Recognition (ICPR), pp. 582–585 (2006)
28. Tsymbal, A.: The problem of concept drift: definitions and related work. Department of Computer Science,Trinity College, Ireland (2004)
29. Uludag, U., Ross, A., Jain, A.: Biometric template selection and update: a case study in fingerprints. Pattern Recogn. 37(7), 1533–1542 (2004)
30. Zhu, X.: Semi-supervised learning literature survey, Technical Report 1530. University of Wisconsin-Madison, Computer Sciences (2005)

Chapter 4
A Hybrid CRF/HMM for One-Shot Gesture Learning

Selma Belgacem, Clement Chatelain and Thierry Paquet

Abstract This chapter deals with the characterization and the recognition of human gestures in videos. We propose a global characterization of gestures that we call the *Gesture Signature*. The gesture signature describes the location, velocity, and orientation of the global motion of a gesture deduced from optical flows. The proposed hybrid CRF/HMM model combines the modelling ability of hidden Markov models and the discriminative ability of conditional random fields. We applied this hybrid system to the recognition of gesture in videos in the context of one-shot learning, where only one sample gesture per class is given to train the system. In this rather extreme context, the proposed framework achieves very interesting performance which suggests its application to other biometric recognition tasks.

4.1 Introduction

A gesture is a short human body motion, in the range of seconds, achieved primarily with arms to generally perform an action. In some situations of disability or constrained environment, the gesture is the only possible mean of communication between humans or between the human being and the machine. In the latter case, the machine identifies gestures using computer vision techniques.

Gesture analysis field includes several themes: characterization, tracking, recognition, segmentation, spotting, etc. As part of our study, we focus on gesture characterization and recognition. Gesture characterization involves extracting information from the data in the aim to discriminate the classes of gestures. Gesture characterization is a necessary step for gestures recognition.

In the case of continuous sign language, recognition must integrate articulated gestures, it must combine segmentation and classification as well. Segmentation consists in determining the limits of gestures in the sequence of video frames. Classification consists in assigning a label belonging to a given vocabulary of gestures to

S. Belgacem · C. Chatelain · T. Paquet (✉)
LITIS EA 4108, University of Rouen, Saint-etienne du Rouvray, France
e-mail: thierry.paquet@univ-rouen.fr

© Springer International Publishing Switzerland 2015
A. Rattani et al. (eds.), *Adaptive Biometric Systems*,
Advances in Computer Vision and Pattern Recognition,
DOI 10.1007/978-3-319-24865-3_4

each sequence of video frames that compose a specific gesture. As stated by Sayre [30], segmentation and classification are two tasks that must be performed simultaneously. The classification task must also integrate knowledge a priori on data such as the vocabulary of gestures, gesture duration, the recording environment, etc. The segmentation step has to face the variability of the duration of gestures, while the classification step has to face the variability of instances of a same gesture.

A gesture is a set of movements performed mainly with hands. It can be represented in a simplified three-dimensional space consisting of its two-dimensional projection and its variation through time. In addition, the recognition system must be robust to recording environment variations. Indeed, the recording conditions are not usually identical between two sequences representing the same gesture. We can observe changes in brightness, backgrounds, colours, objects, etc. Note that the appearance of the involved human may also change (clothes, skin colour, height, etc.).

Markov models are widely applied to the recognition and segmentation of sequential data. They model the temporal dependencies in sequences. They are based on the Markovian assumption that account for the short-term dependencies only, omitting the long-term dependencies in the model.

Although introducing some simplification in the model, generative Markov Models such as hidden Markov models (HMM) [27] allow to introduce a temporal structure between classes that account for high-level knowledge such as a language model. Some other Markov models such as conditional random fields (CRF) [17] are more oriented toward the local discrimination of patterns. In this work, we propose to combine the advantages of these two types of Markov models to provide a hybrid system. We will show that this hybrid system allows the integration of knowledge while being robust to different sources of variability.

Gestures are characterized using an original global description that account for shapes and motions in the video frames. This method describes the location, the velocity and the direction of the motion, based on the optical flow velocity information.

This system was tested using the *"Gesture Challenge 1–2"* dataset proposed by ChaLearn 2011–2012 [11]. The subject of this competition is one-shot gesture learning [11, 40]. We will show later that the lack of training data is another problem that the Markov models are able to solve to a certain extent.

In Sect. 4.2 of this chapter, we present an overview of the gesture recognition applications in the literature, especially the hybrid models combining HMM with other classification methods. In Sect. 4.4, we show the principle of our hybrid model CRF/HMM and explain its interest. Then, in Sect. 4.5, we describe our gesture characterization model. In Sect. 4.4.2.1, we explain how we adapted our hybrid system to the one-shot learning context, in order to cope with the lack of training data. Finally, we will present in Sect. 4.6, the experimental protocol and the evaluation results of our system and its properties.

4.2 Related Works to Gesture Recognition

During the last decade, many studies have been devoted to gesture recognition, and especially in order to design automatic systems that would recognize the sign language. Such systems would allow deaf people to better communicate with machines or with other humans. For example, Vogler and Metaxas[36], Agris et al.[37] and Ong et al.[25] designed a parallel HMM model for signed sentences recognition. They distinguished gesture descriptors such as position, orientation and distance to facilitate the learning process of the HMM and optimize the use of these descriptors. This decomposition is manifested by the generation of one HMM for each descriptor and for each sub-unit of the model.

For gesture sequences recognition, the use of global parallel HMM models is common in the literature[13, 16, 25, 36–38]. HMM models have also been used with a very small number of training examples[13, 16, 38, 40]. This paper addresses the lack of data problem, which is a major problem in the field of machine learning. Konecny et al.[16], Jackson[13] and Weiss[38] proposed a global HMM model for gesture sequences recognition using single-instance learning databases. The global model is a set of left–right interconnected HMM's modelling each gesture. From each state of each HMM, it is possible to remain in that state or to jump to a subsequent internal or external state. In the model proposed by Jackson[13], each frame of the gesture video is represented by a state. This model remains complex due to the large number of states involved.

The idea of combining HMM with other classification scheme is not new. Such hybrid framework is intended to introduce a better discrimination between classes, than generative models can do. One of the first combination scheme was proposed in the 1990s by the integration of neural networks to HMM's[34]. Such combination is prevalent in the literature in various fields. This type of hybrid models was applied to speech recognition[14, 21, 24, 29, 32, 41], handwriting recognition[3, 9, 15, 19, 20, 22, 33] and gesture recognition[6]. HMM models have also been combined with SVM models for handwriting recognition[8] and with dynamic programming methods for gesture recognition[28]. We noticed that the application of these hybrid models to gesture recognition is recent and not much studied in the literature.

To our knowledge, the only work addressing CRF and HMM combination is the work of Soullard et al.[31], based on the work of Gunawardana et al.[10]. In this work, the authors constrain the learning step of a hidden CRF by initialising it with the parameters of a pretrained HMM. This method ensures the convergence of the hidden CRF learning step and shows the difficulty of learning convergence of such models. The idea of our approach is different and is inspired from neuro-Markovian approaches. The principle of these approaches is to replace the HMM data model, consisting of a mixture of Gaussians, by a discriminative model that classifies local observations. This model is traditionally composed of a neural network which provides local a posteriori probabilities of each class associated to each local observation in the sequence. In this work, we propose the use of a CRF in order to perform this discriminative layer. The CRF layer will discriminate local observations and provide local class posteriors to the HMM layer. These local posteriors are then combined during the HMM decoding stage that integrates more global information embedded in the HMM transition model (known as the language model). According to the principle of our hybrid model, the HMM learning step and the CRF learning step are performed separately. Details of the new hybrid model we propose are presented in Sect. 4.4.

4.3 Markovian Models

4.3.1 Hidden Markov Models (HMM)

The Hidden Markov Models (HMM) [2] are probabilistic generative statistical models used for sequence recognition. Their principle is to generate observations based on some hidden states. The joint probability $p(y_{1:T}, x_{1:T})$ (Eq. 4.1) for the observation sequence $x_{1:T}$ and the hidden state sequence $y_{1:T}$ is derived from the particular generative graphical model depicted on Fig. 4.1. This simple graphical model is obtained at the expense of two restrictive assumptions: each observation x_t depends only on the current hidden state Y_T (thus assuming observations to be conditionally independent between each other) and each hidden state Y_T depends only on the previous state y_{t-1} (for an order 1 Markov model). Finally, these assumptions lead to the factorization of Eq. 4.1.

$$p(y_{1:T}, x_{1:T}) = p(y_1)p(x_1|y_1) \prod_{t=2}^{T} p(y_t|y_{t-1})p(x_t|y_t) \qquad (4.1)$$

Through the inference phase, the most likely sequence of hidden states Y^* that describes the given sequence of observations X is determined. Viterbi algorithm [35] is used to find this best sequence.

The graphical data modelling with a HMM model is very interesting. This model is used to guide the decoding process by preserving the structural consistency over time. This model makes it possible to integrate high-level *a priori* knowledge such as syntactical information or duration. Another advantage of HMM's is that they do not require having labelled frames, as the EM-based training process is able to infer local labels from global label given at gesture level.

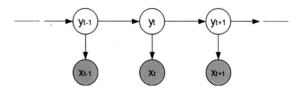

Fig. 4.1 Graphical representation of a HMM: each observation x_t depends only on the hidden state y_t and each hidden state y_t depends only on the previous state y_{t-1}

Generative models such as HMM use Gaussian mixtures to approximate the data distribution. When training data are too few, modelling becomes poor and inadequate which is a major drawback of HMM's. However, discriminative models can remedy this problem. We present in the next section a discriminant sequential Markov model: CRF. This model was proposed by Lafferty et al. [17]. It has some advantages that can address HMM problems.

4.3.2 Conditional Random Fields (CRF)

Conditional random fields (CRF) [17] are discriminative Markov models known for their classification ability. They have been designed in order to model the decision process of labelling a sequence. Therefore, they account for the a posteriori probability of a particular sequence of labels. As depicted in Fig. 4.2, at each time step, a label depends on the previous label (Markov assumption) and may depend on the whole observation sequence X. Making no requirement about the conditional independence of the observation data. The graphical representation of a CRF model is a linear undirected graph with a HMM similar structure. Weights associated to each arc are no longer probabilities but potential functions reflecting the adequacy (or the link) between the two nodes.

The probability of a state sequence $Y = y_{1:T}$ knowing the sequence of observations $X = x_{1:T}$ is computed by:

$$p(Y|X) = \frac{1}{Z(X)} \exp\left(\sum_{t=1}^{T}\sum_{k=1}^{K} \lambda_k f_k(y_{t-1}, y_t, X, t)\right), \qquad (4.2)$$

where $Z(X)$ is a normalization term.

f_k, $\forall k \in [1, K]$ are the feature functions. There are two types of feature functions: feature functions of transitions between successive states representing Markov dependencies and observation feature functions. λ_k is the f_k function weight. The weights λ_k, $\forall k \in [1, K]$ are the parameters to be optimized during the CRF training procedure.

As opposed to HMM, CRF are not able to model high-level information such as a language model or syntactical rules. They are local classifiers in a sequential process. Thus, the high-level knowledge must be introduced in postprocessing as

Fig. 4.2 A representation of the graphical structure of the linear CRF

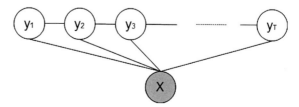

an additional step of filtering in order to guaranty the structural labelling consistency. The HMM's generative framework has this ability of coping with high-level structuring information.

Finally, if we compare the advantages and disadvantages of CRF and HMM, we find a certain complementarity between the two models. Therefore, we propose to combine these two models in a hybrid framework that we present in the next section.

4.4 Hybrid CRF/HMM Model

4.4.1 Overview of the CRF/HMM Model

In this section, we present our hybrid CRF/HMM system for gesture recognition. It combines the discriminative ability of CRF with the modelling ability of HMM. Combining the two models is performed in an easy and straightforward way derived from the literature. The discriminative CRF stage provides local class posterior probabilities that are fed to the HMM stage that account for more global constraints regarding the label sequence. Figure 4.3 shows the proposed hybrid system.

Following this model, the HMM probability $p(y_{1:T}, x_{1:T})$ (see Eq. 4.3) depends on the posteriors computed using the CRF.

$$p(y_{1:T}, x_{1:T}) = p(x_1|y_1)p(y_1) \prod_{t=2}^{T} p(x_t|y_t)p(y_t|y_{t-1}) \qquad (4.3)$$

However, $p(x_t|y_t)$ is a likelihood, while the CRF outputs posteriors $p(y_t|x_t)$. Therefore, $p(x_t|y_t)$ is computed from $p(y_t|x_t)$ using Bayes' rule:

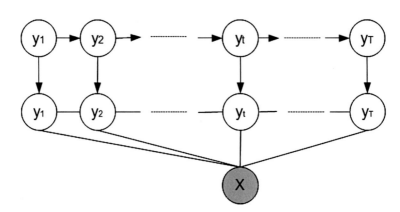

Fig. 4.3 The graphical model CRF/HMM

$$p(x_t|y_t) = \frac{p(y_t|x_t)p(x_t)}{p(y_t)} \tag{4.4}$$

As every gesture class are considered to be equally likely, $p(y_t)$ is a constant $\forall t \in \mathbb{N}$. The aim of the decoding process is to find the state sequence $y_{1:T}$ that maximizes $p(y_{1:T}, x_{1:T})$. As the observation probability $p(x_t)$ is time independent, $p(x_t)$ is not involved in the maximization of $p(x_t|y_t)$. Hence, the maximization of $p(x_t|y_t)$ turns toward the maximization of $p(y_t|x_t)$.

Given that the CRF are able to take into account the whole observation sequence to compute the posteriors of each class, one can state that $p(y_t|x_t) = p(y_t|x_{1:T})$. Let us recall that $y_{1:T}$ and $x_{1:T}$ are noted Y and X.

This is computed within the CRF using the forward-backward algorithm [1], where the forward probability α_t and the backward probability β_t are computed using the following recurrences:

$$\alpha_t(i) = p(x_1x_2 \ldots x_t, y_t = s_i) = \sum_{j=1}^{N_s} \alpha_{t-1}(j)\psi_t(s_i, s_j, o_l), \tag{4.5}$$

$$\beta_t(i) = p(x_{t+1}x_{t+2} \ldots x_T, y_t = s_i) = \sum_{j=1}^{N_s} \beta_{t+1}(j)\psi_{t+1}(s_i, s_j, o_l), \tag{4.6}$$

where

$$\psi_t(s_i, s_j, o_l) = \exp(\sum_{k=1}^{K} \lambda_k f_k(y_t = s_i, y_{t-1} = s_j, x_t = o_l)) \tag{4.7}$$

and s_i, s_j are hidden state that belong to $\mathscr{S}$, and o_l is an observation that belong to $\mathscr{O}$. Finally, following the forward-backward procedure, we have:

$$p(X) = \sum_{j=1}^{N_s} \alpha_T(j) = \sum_{j=1}^{N_s} \beta_1(j) = \sum_{j=1}^{N_s} \alpha_t(j)\beta_t(j), \tag{4.8}$$

$$p(y_t = s_i|X) = \frac{p(y_t = s_i, X)}{p(X)} = \frac{\alpha_t(i)\beta_t(i)}{\sum_{j=1}^{N_s} \alpha_t(j)\beta_t(j)} = \gamma_t(i). \tag{4.9}$$

4.4.2 Training the CRF/HMM Model

We chose to achieve a separated training of HMM and CRF. The HMM training provides the transition matrix between gesture states. Transition models are learned separately for each gesture class and gathered into a global model for decoding gesture sequences. This model is described in Sect. 4.4.4.

As CRF do not benefit from an embedded training stage like HMM, it is necessary to build a frame-labelled learning dataset. This is achieved using the initial HMM model of gesture trained on the dataset that are used in a forced alignment mode that provides the desired frame labelling. Then, the CRF learns a single model for every gestures, considering as many classes in the model as there are sub-gestures. The number of sub-gestures is equal to the number of states in the HMM model of gesture.

4.4.2.1 CRF/HMM Adaptation to One-Shot Learning

In this section, we focus on the learning of the recognition system using a unique sample per class. These learning conditions are interesting since the annotation efforts are extremely reduced in this case. Furthermore, using a single sample per class allows to speed up the learning process.

The one-shot learning framework has been quite extensively used for gesture analysis and recognition [13, 16, 38–40]. These system are generally made up of a standard recognition method that has been adapted to the one-shot learning framework. We now describe the adaptation of our models (HMM and CRF) to one-shot learning.

To model the feature space, the HMM relies on Gaussian mixtures estimated on the learning database. When considering a very reduced number of samples, the Gaussian distribution parameters are very difficult to estimate, especially the variance. Therefore, first we limited the mixture to one Gaussian per gesture class. Second, the variance is computed on every gesture class in order to increase the amount of data and improve the estimation. Doing that, each gesture class has the same variance. Although these two tricks are a limitation of the initial method, the experiments showed the interest of such an adaptation.

In its initial form, the CRF method is mathematically able to deal with either discrete or continuous features; however, since the CRF classification stage is derived from a logistic regression, it is more adapted to discrete features than continuous. This is even more true when the number of samples is small. Therefore, we turned toward the use of a feature quantization procedure. It allows to efficiently tune the parameters linked to each discrete feature value. Notice that some recent developments have introduced hidden CRF models in order to cope with continuous features [26]. But such a framework would require more data than possible in the one-shot learning context.

The quantification is achieved using a uniform scalar quantifier that maps each continuous feature into N_q discrete features, according to the following equation:

$$Q : [-V_{max}, V_{max}] \longrightarrow [-N_q, N_q] \qquad (4.10)$$
$$x \longmapsto \frac{x \times N_q}{V_{max}}.$$

We empirically tuned the value N_q in order to reach the best recognition performance using a validation procedure. We found that $N_q = 16$ was the best value.

4.4.3 Structure and Parametrization of the CRF/HMM Model

As for a standard HMM, the HMM of our hybrid structure is made of states describing each gesture. Although the gesture duration can be modelled through the state autotransitions, it is known that a better modelization can be achieved by setting a variable number of states per gesture. We experimentally checked that this strategy outperforms the performance of the same system with a fixed number of states per gesture. The number of states of each gesture i is determined automatically depending on its frame length $f_g(i)$. The theoretical number of frames per state, denoted f_s, is one hyperparameter of the system. We denote the number of states of a gesture model i; $N_e i = f_g(i)/f_s$. As we already mentioned, we limit the data model to have only one Gaussian per state.

The CRF part of our hybrid model has a standard linear structure, as shown in Fig. 4.3. The CRF training leads to a single model that discriminates all the gestures of the dataset. As explained in the previous section, the CRF formulation allows to consider an observation window, including the current observation and a neighbouring context to be determined. To adapt the system to the gesture duration variability, we chose a variable size f_w of the observation window *CRFwind*. f_w is statically estimated on the learning databases. In order to avoid overfitting the CRF, a regularization term has been empirically tuned to a value of 1.5.

4.4.4 Decoding Using the CRF/HMM Model

The gesture sequence to recognize may contain an arbitrary number of gestures, in an arbitrary order. Therefore, the model should evenly switch between the gesture models. This can be modelled by gathering all the gesture model within a global sequence model, as shown in Fig. 4.4. In this model, each line represents an isolated gesture, with a variable number of state. This global model allows to describe any arbitrary gesture sequence with equiprobable gesture transition probabilities.

Fig. 4.4 The recognition model of gesture sequences using HMM. $e_j^{g_i}$ represents the state j of the gesture i

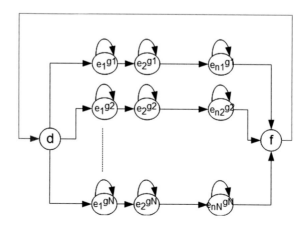

4.5 Global Gestures Characterization

Gestures characterization requires velocity descriptors and shape descriptors as well. Considering that signers can wear clothes in different colours and have different skin colours; colour descriptors are not included in our characterization model.

In this section, we present a set of motion descriptors deduced from optical flows velocities. We call this set of descriptors *Gesture Signature* (GS). We also propose to include shape descriptors extracted with histogram of oriented gradients (HOG). Such descriptors will account for shape descriptors.

4.5.1 Characterization with Optical Flows: Gesture Signature

Optical flows describe local velocities at the pixel level. They are known for their robustness to brightness changes [4]. They are invariant to colours and object distortion. Optical flows are able to describe simultaneously all movements in the scene without any segmentation. Therefore, this method seems adequate to simultaneously extract a maximum of information on body motion, while being robust to variability of colour, shape and brightness. In what follows, we propose a feature vector whose components are combinations of velocity values computed with optical flows.

Hand movements are usually located on the left and the right part of the image, so it is advantageous to divide the image into two vertical sections as shown in Fig. 4.5. Thus, the description of the movement is better localized and motions are characterized in these two distinct regions.

Each part of the image is described by a gesture signature which consists of nine descriptors derived from positive and negative horizontal components V_X^+ and V_X^-, and nine descriptors derived from vertical components V_Y^+ and V_Y^-. These components

Fig. 4.5 The directions of the optical flow components (image from a ChaLearn database video)

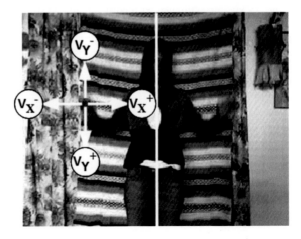

are derived from optical flows at each pixel of the image at position p (Fig. 4.5). Obviously, for each pixel p, two of these four values are null; one pixel can have only one direction according to the x-axis and one direction according to the y-axis.

For a given direction, these nine descriptors consist of four movement *location* descriptors, two movement *velocity* descriptors and three movement *orientation* descriptors. Although these features are simple, they are complementary and describe precisely the gesture changes since location, velocity and orientation are the main components of a gesture.

Table 4.1 shows the 18 features set.

The eight horizontal and vertical location features are related to inertia centre coordinates. They represent the vertical and horizontal positions of velocity centres with respect to the global movement of the considered portion of the image.

There are four features of movement velocity and strength. The first descriptor gives an energy information of the movement. It is inversely proportional to the quadratic mean of the moving pixels velocities. For normalization reasons, we use the inverse of this quadratic mean. The second descriptor gives information about the motion amplitude. It is the median of the moving pixels velocities. The median integrates information about the linear momentum, where the mass is replaced in our case by the number of moving pixels. The median also reduces the noise effect. V_X^* and V_Y^* components are the medians of a threshold velocity vector which is computed with optical flows. Values of the threshold are given below.

$$S_{V_X} = \frac{\sum_{p=1}^{N_{px}^s} |V_X(p)|}{N_{px}^s} \qquad S_{V_Y} = \frac{\sum_{p=1}^{N_{px}^s} |V_Y(p)|}{N_{px}^s}$$

The six movement orientation features are statistics on pixels moving in the same direction, positive or negative. The first two descriptors characterize the amount of pixels moving in the same direction. The third descriptor characterizes the dominant direction of the movement. Those three descriptors characterize the relationship or the symmetry between the two main movement groups whose orientations are opposite. Figure 4.6 shows the interest of these descriptors and illustrates the symmetry information. Thus, by analysing the variation of these three descriptors, we can deduce the type of associated movement. Hence the importance and the complementarity of these three orientation descriptors.

4.5.2 Characterization with HOG

For a complete gesture characterization, we add global contour features extracted with a classic shape descriptor; histograms of oriented gradients (HOG). To apply this descriptor, we resumed the implementation of Dalal et al. [7]. nine directions are used to quantify gradients inclination angles calculated on the image. According to the work of Dalal et al. [7], detecting people with these nine orientations is efficient.

Table 4.1 The eight movement **location** features, the four motion **velocity** features and the six movement **orientation** features

	Descriptor	Horizontally	Vertically
Location	Average Abscissa of pixels moving in the positive direction (AAP)	$\frac{1}{I_w} \times \frac{\sum_{p=1}^{N_{px}^+} |V_X^+(p)|x_p}{\sum_{p=1}^{N_{px}^+} |V_X^+(p)|}$	$\frac{1}{I_w} \times \frac{\sum_{p=1}^{N_{px}^+} |V_Y^+(p)|x_p}{\sum_{p=1}^{N_{px}^+} |V_Y^+(p)|}$
	Average ordinate of pixels moving in the Positive direction (AOP)	$\frac{1}{I_h} \times \frac{\sum_{p=1}^{N_{px}^+} |V_X^+(p)|y_p}{\sum_{p=1}^{N_{px}^+} |V_X^+(p)|}$	$\frac{1}{I_h} \times \frac{\sum_{p=1}^{N_{px}^+} |V_Y^+(p)|y_p}{\sum_{p=1}^{N_{px}^+} |V_Y^+(p)|}$
	Average Abscissa of pixels moving in the negative direction (AAN)	$\frac{1}{I_w} \times \frac{\sum_{p=1}^{N_{px}^-} |V_X^-(p)|x_p}{\sum_{p=1}^{N_{px}^-} |V_X^-(p)|}$	$\frac{1}{I_w} \times \frac{\sum_{p=1}^{N_{px}^-} |V_Y^-(p)|x_p}{\sum_{p=1}^{N_{px}^-} |V_Y^-(p)|}$
	Average ordinate of pixels moving in the negative direction (AON)	$\frac{1}{I_h} \times \frac{\sum_{p=1}^{N_{px}^-} |V_X^-(p)|y_p}{\sum_{p=1}^{N_{px}^-} |V_X^-(p)|}$	$\frac{1}{I_h} \times \frac{\sum_{p=1}^{N_{px}^-} |V_Y^-(p)|y_p}{\sum_{p=1}^{N_{px}^-} |V_Y^-(p)|}$
Velocity	Global velocity inverse (GVI)	$\sqrt{\frac{N_{px}}{\sum_{p=1}^{N_{px}} (V_X(p))^2}}$	$\sqrt{\frac{N_{px}}{\sum_{p=1}^{N_{px}} (V_Y(p))^2}}$
	Maximum velocities median (MVM)	$\frac{1}{S_{V_X}} \times |V_X^*|$	$\frac{1}{S_{V_Y}} \times |V_Y^*|$
Orientation	Proportion of the pixels moving in the positive direction (PPP)	$PPP_X = \frac{N_{px}^{V_X^+}}{N_{px}}$	$PPP_Y = \frac{N_{px}^{V_Y^+}}{N_{px}}$
	Proportion of the pixels moving in the negative direction (PPN)	$PPN_X = \frac{N_{px}^{V_X^-}}{N_{px}}$	$PPN_Y = \frac{N_{px}^{V_Y^-}}{N_{px}}$
	Dominant orientation (DO)	$DO_X = \frac{N_{px}^{V_X^+} - N_{px}^{V_X^-}}{N_{px}}$	$DO_Y = \frac{N_{px}^{V_Y^+} - N_{px}^{V_Y^-}}{N_{px}}$

In HOG, these nine directions are weighted by the corresponding gradients norms in the image computing cells. It applies sliding window self-superposition, generating redundant histograms and a very large HOG feature vector (with size equal to 3780). To alleviate this vector, we used the average operator on two levels. The first simplification level is based on the HOG visualization algorithm proposed by Jurgen Brauer.[1] The idea of this algorithm is to average the redundant histograms on image

---

[1] http://www.juergenwiki.de/work/wiki/doku.php?id=public%3ahog_descriptor_computation_and_visualization.

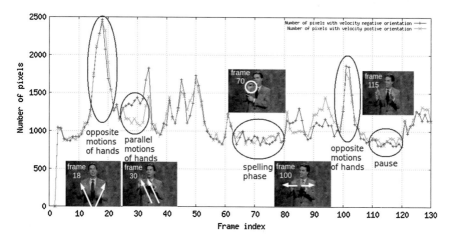

Fig. 4.6 Evolution of the descriptors PPP_X and PPN_X in a video from SignStream database [23]. Two *curves* superimposed with a presence of a peak correspond to an opposite movement of the two hands. A strong difference between the two curves corresponds to a parallel movement of both hands in the dominant direction. A stagnation of the two curves corresponds to fixed hands (frame 70)

cells keeping nine gradient directions in each cell. The second level of the HOG descriptor simplification, applied in our case, is to average the gradient amplitudes on larger image blocks that we call meta-blocks. We partitioned the image blocks to meta-blocks. For each meta-block and for each orientation, we compute mean amplitude in all meta-block cells. Tests were carried out using 4 and 16 meta-blocks. We obtain then nine amplitude averages per meta-block which leads to a descriptor of size $9 \times 4 = 36$ or $9 \times 16 = 144$. In our case, HOG are computed on the difference of two successive images in order to characterize only moving patterns.

4.6 Experimental Protocol

In this section, we explain the experimental protocol: databases, evaluation methods, feature vector variants and implementation tools.

4.6.1 Databases

Our recognition system has been evaluated on public databases designed for the ChaLearn 2011–2012 competition [11]. We did not participate to this competition but we were able to compare our system to those of the participants thanks to the evaluation platform proposed by the competition organizers.[2] We detail the results of this evaluation in Sect. 4.7.

ChaLearn databases are made of three types of resources: 480 system development sub-databases named *devel*, 20 system validation sub-databases named *valid* and 40 system final evaluation sub-databases named *final*. The 1–20 *final* sub-databases were tested in the first round of the competition and 21–40 *final* sub-databases were tested in the second round of the competition. This final evaluation classifies participants in the ChaLearn competition.

Each of these sub-databases contains 47 pairs of videos. Each video pair presents the same scene in two formats: RGB colour format and depth format. These videos are recorded using a Kinect (TM) camera at a frequency of 10 frames per second, with a resolution of 240 × 320 pixels. Videos of the same sub-database share the same scenic features: same actor, same background, same recording conditions, same theme and same gesture vocabulary. However, these scenic characteristics vary from sub-database to another. 20 players participated in the making of these databases, one actor per sub-database. These databases present 30 vocabularies composed of 8–15 gestures belonging to various themes such as video games, distance education, robot control, sign language, etc.

Each sub-database includes two sets of video: a training set $\mathbb{G}$ and a test set $\mathbb{S}$. The training set $\mathbb{G}$ consists of 10 videos. Each video contains a single and isolated instance of a gesture: *one-shot learning* databases. The test set $\mathbb{S}$ consists of 40 videos. Each video includes a sequence of 1–5 successive gestures separated by a common break point. Gestures organization in each test sequences is random, there is no specific gestures grammar.

We summarize in the following subsection the various feature vectors used for the tests.

4.6.2 Feature Vector Variants

Table 4.2 presents the different variants of the feature vector $\mathbf{c}$ we used in our experiments. We index each variant by its size $l(\mathbf{c})$. $l(\mathbf{v}(GS))$ is the number of gesture signature features. $l(\mathbf{v}(HOG))$ is the number of HOG features. Some variants of the feature vector $\mathbf{c}$ are applied to two data formats (RGB image and depth image).

4.6.3 Evaluation Metric

The organizers of the ChaLearn competition defined a global evaluation metric on all test sequences based on the Levenshtein distance, also called edit distance [18]. This form of global error is denoted by $\mathscr{L}_{ch}$ and given by Eq. 4.11.

---

[2]https://www.kaggle.com/c/GestureChallenge2.

Table 4.2 Feature vector variants adopted in the experiments

Total size $l(\mathbf{c})$	Descriptor				
	Gesture signature GS			HOG	
	$l(\mathbf{c}(GS))$	Description		$l(\mathbf{c}(HOG))$	Description
54	18	No image division		36	4 meta-blocks
52	16	No median, no image division		36	4 meta-blocks
180	36	Image division into 2 parts		144	16 meta-blocks
360	72	Image division into 2 parts, 2 data formats		288	16 meta-blocks, 2 data formats
72	72	Image division into 2 parts, 2 data formats		0	HOG not applied

$$\mathscr{L}_{\text{ch}} : \mathbb{D} \longrightarrow \mathbb{R} \qquad\qquad (4.11)$$
$$\mathbb{S} \longmapsto \frac{\sum_{s\in\mathbb{S}} L(\mathscr{R}(s), \mathscr{T}(s))}{\sum_{s\in\mathbb{S}} l(\mathscr{T}(s))},$$

where $\mathbb{D}$ is the set of test databases, $\mathbb{S}$ is the set of test sequences, s is the sequence of gestures, $\mathscr{R}(s)$ is the system recognition result of sequence s, $\mathscr{T}$ is a function giving the ground truth sequence s, $L(., .)$ is the Levenshtein distance and $l(v)$ gives the size of a vector v.

We use the ChaLearn form of the error $\mathscr{L}_{\text{ch}}$ to compare our recognition system to ChaLearn participants recognition systems. However, let us emphasize that $\mathscr{L}_{\text{ch}}$ is slightly different from the classical Levenshtein distance (see Eq. 4.12), which is bounded and seems more generic. Thus, to present the main results of our various tests, we use the classic error form.

$$\mathscr{L} : \mathbb{D} \longrightarrow [0, 1] \qquad\qquad (4.12)$$
$$\mathbb{S} \longmapsto \frac{1}{|\mathbb{S}|} \sum_{s\in\mathbb{S}} \frac{L(\mathscr{R}(s), \mathscr{T}(s))}{l(\mathscr{R}(s)) + l(\mathscr{T}(s))}$$

4.6.4 Implementation Tools

We used the OpenCV library [5] to develop image and video processing methods. HMM gesture recognition methods have been implemented thanks to Torch library,[3] while CRF gesture recognition methods rely on the CRF++ library.[4]

---

[3] http://torch.ch/torch3/.
[4] http://crfpp.googlecode.com/svn/trunk/doc/index.html.

4.7 Gesture Recognition Results

In this section, we present the results of our system, using different variants. We first evaluate the effect of the quantification of continuous features for a discrete CRF in Sect. 7.1. Then, we demonstrate the robustness of the hybrid model CRF/HMM with respect to the number of states and to the various feature vectors in Sect. 7.2. Finally, we compare the recognition results of the hybrid system CRF/HMM to the classic and adapted versions of HMM and CRF in Sect. 7.3. We conclude this section by presenting our rank compared to participants at the ChaLearn competition.

All recognition performance results of the hybrid system CRF/HMM presented in this section are obtained with tests performed with an adapted CRF/HMM as explained in Sect. 4.4.2.1 unless otherwise stated. Adapted HMM and adapted CRF recognition systems cited in this section are also adapted as explained in Sect. 4.4.2.1.

4.7.1 Evaluation of the Features Quantization for CRF

Although CRF are able to cope with continuous features, it has been shown that discretizing the feature set could increase its performance, especially when the number of training examples is small [12].

Indeed, continuous CRF put a single weight for all values of a descriptor. Whereas a reduced value of this descriptor does not necessarily mean that it has no importance and a high value of this descriptor does not mean that it is really important. This way of managing weights can be suitable to weight a score function whose values have a monotonous importance. However, for a descriptor, distinctive ranges of values can change from one descriptor to another. Thus, discrete CRF, which give a distinct weight for each discrete feature value, provide more specification to features, which subsequently increases the discrimination of classes. Therefore, discrete CRF seems an adequate model for one-shot learning case, as we noted in the Sect. 4.4.2.1.

Figure 4.7 (left) presents the CRF/HMM recognition performance in both continuous and discrete characteristics cases by varying the number of frames per state for the HMM component. Discrete system performances clearly outperform continuous system performances, which demonstrate the interest of quantification. Let us also mention that the learning time of continuous CRF (estimated in hours) largely exceeds the learning time of discrete CRF (estimated in minutes). This is another advantage of discrete CRF.

4.7.2 Robustness of the CRF/HMM Approach

In this section, we analyse the influence of several parameters on our CRF/HMM approach results: number of frames per states, gesture duration and feature vector.

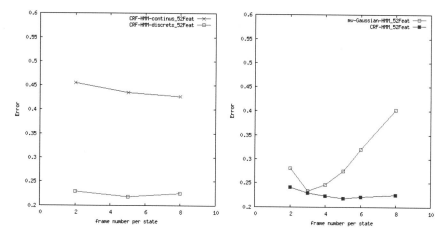

Fig. 4.7 *Left* CRF/HMM gesture recognition results with continuous and discrete component. *Right* CRF/HMM and adapted HMM systems robustness to the variation of the number of frames per state

4.7.2.1 Robustness to Changes in the Number of Frames per State

Figure 4.7 (right) shows the recognition error $\mathscr{L}$ of adapted HMM and CRF/HMM systems with respect to the number of frame per state f_s. For each value of f_s, the recognition system is re-learned. One can observe that the CRF/HMM system outperforms HMM, and that the CRF/HMM system provides extremely stable results, while the system performance of HMM is strongly variable. This is an interesting feature since it does not require a fine hyperparameter tuning for reaching good results.

4.7.2.2 Robustness to Changes in the Gesture Duration

The change in the number of frames per state has a direct impact on the CRF/HMM robustness to the gesture duration variation. With a large number of frames per state, CRF/HMM system is able to handle the temporal elasticity of a gesture. In other words, when a gesture expands or narrows through the number of frames in the test data, CRF/HMM system is able to align the gesture model on the data and decode them. In addition, CRF component are able to implicitly manage narrowing and expansion of data through their local decision which is independent from the data global model, unlike HMM which are dependent on a graph-oriented model without jumps. Thus, to manage the temporal elasticity of gestures, a simple structure of the hybrid model with a reduced number of states can replace a complex HMM system with jumps between states and a complete connection as adopted by some participants of the ChaLearn competition [13, 16, 38].

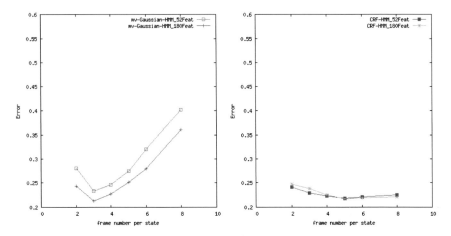

Fig. 4.8 Adapted HMM (*left*) and CRF/HMM (*right*) robustness to the variation of the feature vector

4.7.2.3 Robustness to Changes in the Feature Vector

Figure 4.8 present the variation of the error $\mathscr{L}$ in terms of the number of frames per state f_s for two HMM systems (left) and for two CRF/HMM systems (right). Each pair of systems is evaluated with two different feature vectors. When the feature vector size decreases, CRF/HMM keep almost the same performance. In other words, a minimum of features is sufficient for CRF HMM, whereas for classic HMM, features addition increases greatly the recognition performance. This recognition ability with a reduced number of features makes features extraction task easier and faster.

These three CRF/HMM robustness property prove that with a simple system, it is possible to reach high recognition performance thanks to CRF and HMM advantages combination and disadvantages compensation. We can see the simplicity of the CRF/HMM system at three levels: (a) a simple model structure with a reduced number of state without jumps nor complete connection; (b) a reduced number of features; and (c) a training dataset reduced to an example by class.

4.7.3 Evaluation of the CRF/HMM Using the ChaLearn Platform

We present in this subsection the recognition results of our best hybrid system CRF/HMM on the *valid* and *final* databases, as well as our ranking in the ChaLearn competition.

We first present a comparison of the performance of the main recognition systems that we studied (Table 4.3) on the *devel* databases. The 52 feature vector has been

Table 4.3 The recognition results of various recognition systems based on HMM and CRF and tested on 20 *devel* databases

System	$l(\mathbf{c})$	f_s	Error: $\mathscr{L}$
classic HMM	52	6	0.36
adapted HMM	52	3	0.23
classic CRF (continuous)	52	$f_g(g)$	0.29
adapted CRF (discrete)	52	$f_g(g)$	0.28
CRF/HMM (adapted)	52	5	**0.22**

Table 4.4 The recognition results of our best hybrid system CRF/HMM on 20 *valid* databases, 20 *final* 1–20 databases and 20 *final* 21–40 databases (each database category contains about 750 total sequences test in the order of 200 frames each)

Database category	Error		Ranking
	$\mathscr{L}$	$\mathscr{L}_{ch}$	
Valid	0.177193	0.348812	–
Final 1–20 (1st round)	0.147924	0.296440	7th
Final 21–40 (2nd round)	0.122398	0.252357	7th

chosen since it provides good results while keeping a compact representation (see Table 4.2). It is identical for all the systems. The number of frames per state f_s has been optimized for each system. $f_g(g)$ represents the size of the learned gesture, which means that every gesture is represented by a single class, subclasses that correspond to states in the case of HMM do not exist in the case of CRF. On the other hand, a postprocessing step is applied to the classic and adapted CRF in order to filter their recognition results. Without this step recognition error exceeds 0.5. Table 4.3 shows that the performance of the proposed hybrid system CRF/HMM clearly outperforms the recognition performances of other systems.

In order to rank our system in the ChaLearn 2011–2012 competition, we tested the hybrid system on *valid* and *final* databases provided during the competition. Table 4.4 shows the hybrid system CRF/HMM recognition error values computed with both evaluation methods $\mathscr{L}$ and $\mathscr{L}_{ch}$ on *valid* and *final* databases. Table 4.4 presents the CRF/HMM system rank on both database categories using the $\mathscr{L}_{ch}$ error. It appears that we ranked at the 7th position among 559 systems from 48 participants for both first and second rounds. The complete list with their score (the $\mathscr{L}_{ch}$ error) is available on the Kaggle website for the first[5] and the second round.[6] We achieved this rank using only RGB format data.

[5]https://www.kaggle.com/c/GestureChallenge/leaderboard.
[6]https://www.kaggle.com/c/GestureChallenge2/leaderboard.

Beside the competition, for a data size equal to 750, we demonstrated with the statistical unilateral student test that our hybrid model CRF/HMM significantly outperforms classic models HMM and CRF. CRF/HMM also outperforms the adapted HMM[7] with a confidence level of 99 % and the adapted CRF (see footnote 7) with a confidence level of 99.5 %.

These results and this study show that the CRF/HMM hybrid system is a system that has better performance than other classic systems (HMM and CRF), is robust to different variations, and is interesting and practical in the real-world problem such as one-shot learning.

4.8 Conclusion

In this chapter, we proposed a new hybrid system for gesture recognition CRF HMM. We demonstrated that this combination of Markov models benefits from each model advantages without undergoing its drawbacks. These Markovian models have been adapted to one-shot learning context in order to improve their recognition ability. We also proposed a new gesture characterization model which is a gesture Signature based on optical flows.

We demonstrated that these gesture characterization and recognition models constitute a robust recognition hybrid system that opens up new perspectives for sequential Markov models. An interesting perspective of our gesture recognition work concerns the gesture detection task, called the gesture *spotting*. Gesture spotting consists on locating and labelling specific gestures in videos. It can be applied in video documents management contexts such as video retrieval, categorization and indexing. Our recognition model could be adapted to the spotting task by representing false examples through an additional class to the gestures vocabulary.

Finally, we demonstrated in this chapter Markov systems ability to model and manage spatio-temporal variations of sequential data, including gestures. The modelling evolution of the human activity contributes to the evolution of computer vision techniques and subsequently contributes to the evolution of human–machine interaction systems.

References

1. Austin, S., Schwartz, R., Placeway, P.: The forward-backward search algorithm. In IEEE ICASSP, pp. 697–700 (1991)
2. Baum, L.E., Petrie, T.: Statistical inference for probabilistic functions of finite state Markov chains. Ann. Math. Stat. **37**, 1554–1563 (1966)
3. Bengio, Y., LeCun, Y., Nohl, C., Burges, C.: LeRec: a NN/HMM hybrid for on-line handwriting recognition. neural Comput. **7**(6), 1289–1303 (1995)

[7]The mentioned adaptation is the model adaptation to the one-shot learning context.

4. Bhandarkar, Suchendra M., Luo, Xingzhi: Integrated detection and tracking of multiple faces using particle filtering and optical flow-based elastic matching. CVIU **113**(6), 708–725 (2009)
5. Bradski, Gary, Kaehler, Adrian: Learning OpenCV: Computer Vision with the OpenCV Library. O'Reilly, Cambridge (2008)
6. Corradini, A.: Real-time gesture recognition by means of hybrid recognizers. In: Gesture Workshop, vol. 2298, pp. 34–46. Springer (2001)
7. Dalal, N., Triggs, B.: Histograms of oriented gradients for human detection. In: International Conference on Computer Vision & Pattern Recognition, vol. 2, pp. 886–893 (2005)
8. Ganapathiraju, A., Hamaker, J., Picone, J.: Hybrid SVM/HMM architectures for speech recognition. In: INTERSPEECH, ISCA, pp. 504–507 (2000)
9. Gilloux, M., Lemarie, B., Leroux, M.: A hybrid RBF network/hidden Markov model handwritten word recognition system. In: ICDAR, pp. 394–397 (1995)
10. Gunawardana, A., Mahajan, M., Acero, A., Platt, J.C.: Hidden conditional random fields for phone classification. In INTERSPEECH, ISCA, pp. 1117–1120 (2005)
11. Guyon, I., Athitsos, V., Jangyodsuk, B., Hamner, P., Escalante, H.: ChaLearn gesture chalienge: Design and first results. In CVPR Workshops, pp. 1–6. IEEE (2012)
12. Hebert, D., T. Paquet, Nicolas, S.: Continuous CRF with multi-scale quantization feature functions application to structure extraction in old newspaper. In: ICDAR, pp. 493–497 (2011)
13. Jackson, E.: An HMM-based approach for gesture recognition using edge features. In: CVPR 2012 Workshop on Gesture Recognition (2012)
14. Johansen, F.T.: A comparison of hybrid HMM architectures using global discriminative training. In: ICSLP, ISCA (1996)
15. Knerr, S., Augustin, E.: A neural network-hidden markov model hybrid for cursive word recognition. ICPR **2**, 1518–1520 (1998)
16. Konencny, J., Hagara, M.: One-shot learning gesture recognition using HOG/HOF features. In: ICPR 2012 Workshop on Gesture Recognition (2012)
17. Lafferty, J., McCallum, A., Pereira, F.: Conditional random fields: probabilistic models for segmenting and labeling sequence data. In: ICML, pp. 282–289 (2001)
18. Levenshtein, V.I.: Binary codes capable of correcting deletions, insertions and reversals. Soviet Physics Doklady **10**, 707 (1966)
19. Marukatat, S., Artieres, T., Gallinari, P., Dorizzi, B.: Sentence recognition through hybrid neuro-Markovian modeling. In: ICDAR, pp. 731–737 (2001)
20. Matan, O., Burges, C., Lecun, Y., Denker, J.S.: Multi-digit recognition using a space displacement neural network. In: Advances in Neural Information Processing Systems, vol. 4, pp. 488–495 (1992)
21. Morgan, N., Bourlard, H., Renls, S., Cohen, M., Franco, H.: Hybrid neural network/hidden Markov model systems for continuous speech recognition. IJPRAI **7**(4), 899–916 (1993)
22. Morita, M.E., Sabourin, R., Bortolozzi, F., Suen, C.Y.: Segmentation and recognition of handwritten dates: an HMM-MLP hybrid approach. IJDAR **6**(4), 248–262 (2003)
23. Neidle, Carol, Sclaroff, Stan, Athitsos, Vassilis: SignStream: a tool for linguistic and computer vision research on visual-gestural language data. Behav. Res. Methods Instrum. Comput. **33**(3), 311–320 (2001)
24. Niles, L.T., Silverman, H.F.: Combining hidden Markov models and neural network classifiers. In: ICASSP, pp. 417–420 (1990)
25. Ong, S.C.W., Ranganath, S.: Deciphering gestures with layered meanings and signer adaptation. In: IEEE International Conference on Automatic Face and Gesture Recognition, p. 559 (2004)
26. Quattoni, A., Wang, S., Morency, L., Collins, M., Darrell, T.: Hidden conditional random fields. IEEE Trans. Pattern Anal. Mach. Intell. **29**(10), 1848–1852 (2007)
27. Rabiner, L.R.: A tutorial on hidden markov models and selected applications in speech recognition. In: Proceedings of the IEEE, pp. 257–286 (1989)
28. Rajko, S., Qian, G.: A Hybrid HMM/DPA adaptive gesture recognition method. In: ISVC, vol. 3804, pp. 227–234 (2005)
29. Rigoll, G.: Maximum mutual information neural networks for hybrid connectionist-HMM speech recognition systems. IEEE Trans. Speech Audio Process. **2**(1), 175–184 (1994)

30. Sayre, Kenneth M.: Machine recognition of handwritten words: a project report. Pattern Recogn. **5**(3), 213–228 (1973)
31. Soullard, Y.: Hybrid HMM and HCRF model for sequence classification. Bruges, Belgium (2011)
32. Tebelskis, J., Waibel, A., Petek, B., Schmidbauer, O.: Continuous speech recognition by linked predictive neural networks. In: NIPS, pp. 199–205 (1990)
33. Thomas, S., Chatelain, C., Heutte, L., Paquet, T., Kessentini, Y.: A deep HMM model for multiple keywords spotting in handwritten documents. Accepted in Pattern Anal. Appl. (2015)
34. Trentin, E.: A survey of hybrid ANN/HMM models for automatic speech recognition. Neurocomputing **1–4**, 91–126 (2001)
35. Viterbi, A.: Error bounds for convolutional codes and an asymptotically optimum decoding algorithm. IEEE Trans. Inf. Theory **13**(2), 260–269 (1967)
36. Vogler, C., Metaxas, D.: A framework for recognizing the simultaneous aspects of American Sign Language. Comput. Vis. Image Underst. **81**, 358–384 (2001)
37. von Agris, U., Zieren, J., Canzler, U., Bauer, B., Kraiss, K.-F.: Recent developments in visual sign language recognition. Univ. Access Inf. Soc. **6**(4), 323–362 (2008)
38. Weiss, D.: HMM based one shot gesture recognition. In: CVPR 2012 Workshop on Gesture Recognition (2012)
39. Wu, D., Zhu, F., Shao, L.: One shot learning gesture recognition from RGBD images. In: CVPR, IEEE, pp. 7–12 (2012)
40. Yang, Yang, Saleemi, I., Shah, M.: Discovering motion primitives for unsupervised grouping and one-shot learning of human actions, gestures, and expressions. IEEE Trans. Pattern Anal. Mach. Intell. **35**(7), 1635–1648 (2013)
41. Zavaliagkos, G., Austin, S., Makhoul, J., Schwartz, R.M.: A hybrid continuous speech recognition system using segmental neural nets with hidden Markov models. IJPRAI **7**(4), 949–963 (1993)

Chapter 5
An Online Learning-Based Adaptive Biometric System

A. Das, R. Kunwar, U. Pal, M.A. Ferrer and M. Blumenstein

Abstract In the last decade, adaptive biometrics has become an emerging field of research. Considering the fact that limited work has been undertaken on adaptive biometrics using machine learning techniques, in this chapter we list and discuss a few out of many potential learning techniques that can be applied to build an adaptive biometric system. In order to illustrate the efficacy of one of the incremental learning techniques from the literature, we built an adaptive biometric system. For experimentation, we have used multi-modal ocular (sclera and iris) data. The preliminary results have been reported in the results section, which are very promising.

5.1 Introduction

Since the last few decades, intensive research work has been performed in the biometrics arena. Although various accurate systems have been proposed, surprisingly the adaptiveness of such systems to environmental changing conditions or change

---

*The first and the second author have equal contribution in this work.

A. Das (✉) · R. Kunwar · M. Blumenstein
Institute for Integrated and Intelligent Systems, Griffith University,
Queensland, Australia
e-mail: abhijit.das@griffithuni.edu.au

R. Kunwar
e-mail: kunwar.rituraj@gmail.com

M. Blumenstein
e-mail: m.blumenstein@griffith.edu.au

U. Pal
Computer Vision and Pattern Recognition Unit, Indian Statistical Institute,
Kolkata, India
e-mail: umapada@isical.ac.in

M.A. Ferrer
IDeTIC, University of Las Palmas de Gran Canaria, Las Palmas, Spain
e-mail: mferrer@dsc.ulpgc.es

© Springer International Publishing Switzerland 2015
A. Rattani et al. (eds.), *Adaptive Biometric Systems*,
Advances in Computer Vision and Pattern Recognition,
DOI 10.1007/978-3-319-24865-3_5

in biometric traits is low. Change in biometric traits or variation in the traits is a big challenge for the biometrics domain as it can lead to misidentification. The main reasons for misidentification or rejection of the correct individual by biometric systems are scarcity of training samples, the presence of substantial intra-class variation during testing and lack of standardization of the data acquisition environment. Biometric characteristics can change either temporarily or permanently, perhaps due to ageing, diseases or treatment to diseases. In order to handle these issues, it would be preferable to have a biometric system which adapts well to the changing problem. These drawbacks of the present biometric systems have stimulated the interest of researchers in this domain. Recent developments in the adaptive system research area have opened up a new research field and that is "Adaptive Biometrics".

The ideal case of an adaptive biometric system is expected to handle the intra-class variations which changes with time (in many cases). These changes can happen for various reasons like ageing, variations in pose and lack of standardization of data acquisition rules. The advantages of such a system are: learner need not get trained from scratch every time new data is available (as the learning happens continuously with time) and no need to store old data. This aspect of learning will significantly reduce the maintenance cost of biometric system. These are the characteristics which makes this research area so attractive and suitable for real-time scenario.

The existing-automated adaptive biometric systems have adopted semi-supervised learning to create an adaptive system. A semi-supervised learning or online learning semi-supervised learning is a machine learning scheme which uses both labelled and unlabelled data. In such a machine learning systems, the input samples are assigned labels using existing references and the positively classified samples are incorporated into existing references improve the adaptability of the system. The commonly adopted adaptation procedure is to augment the reference set with the newly classified input samples. The efficacy of such systems can be estimated by comparing the obtained performance gain with traditional biometric schemes which do not have any adaptation mechanism. The performance gain of such systems depends on the accurate labelling of the input samples. This is because misclassifications will introduce impostor samples into the updated reference set; the result of which can be counterproductive and the inter-class gap may get affected dramatically.

An adaptive biometric system can also operate in supervised or offline mode, in which biometric samples are manually labelled and updated. The supervised method represents the best-case performance as all the available positive (genuine) samples are used for adaptation. However, manual intervention makes this process time-consuming and expensive. Therefore, it is generally infeasible to manually update references regularly. Despite of these above highlighted advantages, a large amount of limitations are also associated with the existing offline adaptive biometric systems. First of all capturing substantial amount of samples for such systems is quite time-consuming. Commonly adopted self-updating system captures only limited amount of available samples during enrollment. As a result, a large number of input samples with informative and significant variations remain unenrolled and consequently results in limited performance gain over the baseline system. Second present adaptive biometric systems are vulnerable to impostor attacks or spoofing. As a consequence, such system bears the risk of getting adopted by imposter samples. Such consequence can affect the performance of the system dramatically. Even distinguishing between informative, redundant and noisy input samples are not possible in such system, not even in instances of supervised system. Hence as a result, these occurrences can not only affect the performance of the system but also affect the inter-class variance space of the system.

In this chapter, we discuss several existing online/adaptive learning techniques which can be applied in the biometric domain to build an adaptive biometric system which will overcome above discussed drawbacks. And this will bridge the existing gap in the literature. In order to justify the applicability of the discussed existing online theories theory, multi-modal ocular (sclera and iris) biometric trait was used for experimentation with one of the existing techniques.

Ocular biometrics has gained popularity due to the significant progress made in iris recognition. Various reasons advocate this trend: these patterns possess a high degree of randomness, which are never same for any two individuals, not even for identical twins, and this makes it ideal for personal identification. Further the patterns remain stable throughout a person's lifetime; these patterns even differ for the right and the left eye of the same individual. Therefore, it comes to no surprise that iris has been so popular among the commercially available biometric systems. However, unfortunately iris recognition is unfavorably influenced by the frontal gaze direction of the eye with respect to the acquisition device. In such scenario, additional eye trait can be used to mitigate the motion bottleneck of iris biometrics. Among the various other ocular traits available (retinal scan, peri-ocular and sclera), the sclera (the white of the eye) is believed to be the most promising one, and it may be of great significance in improving biometric applicability of iris recognition. Iris patterns are better discerned in the near infrared spectrum (NIR), while vasculature patterns are better discerned in the visible spectrum (RGB). So, the multi-modal ocular biometric using iris and sclera is proposed in visible spectrum. An additional advantage of this multi-modal biometric is that the sensor can sense the biometric trait from a distance hence such systems are hygienic as well. Although this proposed multi-modal biometric has achieved a high-recognition accuracy but their adaption with the environmental condition is an open research issue. Additionally, adaption with respect to occlusion and various gaze and angle of eye is another challenge faced by this biometrics.

Hence, in this chapter, we discuss many existing online/adaptive learning techniques which can be useful to underpin an adaptive biometric system. In particular, it can come to a great rescue for the biometric problem like ocular biometric which is extremely sensitive to various changes as discussed above. Therefore, we applied one of the existing techniques to illustrate its effectiveness in the ocular biometric domain. The organization of the chapter is as follows: Sect. 5.2 highlights the existing works in the literature of adaptive biometrics and followed by few advance work on ocular biometrics. In Sect. 5.3, we discuss various incremental/adaptive learning techniques available in literature including some experimental results, and Sect. 5.4 draws the overall conclusion and future scope of the theory proposed.

5.2 Adaptive Biometrics Literature

In the last decade, adaptive biometric systems have been adequately investigated. Various research approaches have been explored by different researchers. To the best of our knowledge, the first adaptive biometric approach was introduced in [1]. The existing literature of adaptive biometric systems can be categorized by the key attributes of machine intelligence methods proposed or used. Supervised [2–6] against semi-supervised training [3, 7, 8], self-train [8–10] or co-train [3, 11], and online [8, 9] and offline [3, 12, 13].

A general study has been done in regard to adaptive biometric in [14] to address the intuitive questions like:

1. Whether supervised adaptation better than semi-supervised?
2. Whether co-training can outperform self-training?
3. Whether offline adaptation is better than online?

Interesting analysis was performed further to validate the hypothesis [14].

In [2], a two-stage classifier selection technique was proposed for automatically updating the biometric templates. In this approach, a labelling scheme based on probabilistic semi-supervised learning was employed. Soft probabilistic labels were assigned to each batch of input samples by calculating the minimum energy function on the graphical representation. The harmonic function used was unique and ensured that labels were assigned to input sample using both the enrolled and nearby input data. Then the genuinely classified samples undergo the selection process based on risk minimization. The experiment was validated on a DIEE finger print dataset, an appreciable result was achieved.

In [3], effect of different threshold settings were employed for template update and novel solutions was proposed for by passing the threshold selection step. This work analysed and inferred that template update method is better for group specific updating due to presence of different type of population. Efforts have been made in the work to give a preliminary guideline on the type of update procedures that could be followed for a specific group of population. A protocol for simulating real-world situation has also been proposed for the unbiased evaluations of update methods.

Most of the adaptive biometric work in literature deals with finger prints, voice or facial traits. Ocular biometric is one the most accurate and reliable biometric; it offers various biometric traits. Adaptiveness of ocular biometrics is yet to be addressed in the literature. Among biometric traits face [15], vain pattern [16], ocular biometrics like iris [17, 18] is the most promising one. Apart from iris, the human eye has an ocular white surface known as the sclera which contains a texture pattern due to the presence of blood vessels on its surface. The sclera patterns can be acquired easily along with iris in single camera shot and it is visible, even in off-angle eye shot. Therefore, by utilizing the texture pattern of sclera in addition to iris, the performance of an iris recognition system can be significantly improved even with a non-ideal or an off-angle eye data sample.

The first recognized work on sclera biometrics is recorded in [19]. In this paper, the authors discuss methods for conjunctival image preprocessing by computing a

Gaussian filters and Hessian matrix. In order to derive a suitable vascular template for biometric authentication, feature extraction and classification are performed by a minutia based template matching. Here a dataset was built to establish the experiment. In [20], the first automatic sclera detection technique was proposed by a time adaptive active contour-based method. First, automatic segmentation processes for sclera biometric was proposed in [21], Otsu's algorithm was used to segment the sclera from a grey image. Sclera segmentation based on colour image was first proposed in [22], HSV model was used to segment sclera from a colour eye image, and a bank of Gabor filter was used to enhance the sclera. Many features like LBP [7, 23], and GMCL [24] are used in the literature for sclera feature representation. First multi-angled sclera recognition was proposed in [25], further in [26] multi-angled sclera recognition was proposed using multispectral imagery. Few advance work on sclera is found using features like in SIFT [27], OLBP [28], and Dense-SIFT [29]. In [29], K-means followed by Spatial Pyramid matching [30] was used to enhance the feature set. In [31], a feature-based adaptive sclera biometric is proposed. In [32], a liveness-based sclera biometric is proposed.

First, multi-modal eye recognition techniques using sclera and iris was proposed in [33]. Here a score fusion-based technique was adopted to combine the sclera and the iris feature. Further in [22], a quality fusion technique was used to combine sclera and iris feature, and in [34], feature level combination was used to establish sclera and iris based multi-modal biometrics. A survey on sclera recognition is recorded in [35].

To this date, adaptive biometric is relatively less studied and little is known regarding its usefulness. So, the state of the art related to it is not matured yet. Therefore, in the next section, we discuss various generic adaptive learning methods that can be used in adaptive biometrics domain. We also implement one of the adaptive/online learning methods to demonstrate the efficacy of the used online learning method in multi-modal ocular (sclera and iris) biometric domain.

5.3 Generic Adaptive Learning Methods

In this chapter, we address the potential incremental learning techniques which can be applied in the biometric domain to create an adaptive learning system. The motivation to create an adaptive learning system for biometrics is explained below.

In real-world scenarios, where we use machine learning algorithms, we often have to deal with cases where the input data changes its nature with time. In order to maintain the accuracy of the learning algorithm, we frequently have to retrain our learning system, thereby making the system inconvenient and unreliable. This problem can be solved using learning algorithms which can learn continuously with time (incremental/online learning). In contrast, offline learning works fine in an ideal scenario where there is no change in the underlying distribution of the input with time. However, for various reasons this does not often hold in real-time problems that we intend to address (i.e. of robust biometric system) using machine learning.

In contrast to offline learning, ideally, incremental/online learning can be simultaneously trained and tested. Precisely, it needs not stop performing its task (i.e. prediction or classification) if the learner has to update its learning parameters. Learning parameters can be updated as soon as the new training data is available. This leads to the creation of a never ending learning process which can adjust itself even if the environment changes and can perform learning while performing task.

The critical assumption on which most of the incremental learning algorithms are based upon is that previous data is completely or partially accessible. Based on this assumption, to handle streaming data they apply the time windowing technique of either fixed or variable size [36–38]. Others have handled streaming data by weighting models in the ensemble [39–41] or by weighting the data [42] or by retaining only the relevant subset of previous data [43, 44]. We assume for our experimentation that we have no access to the previous data, thus making the algorithm capable to handle the scenario where old data is inaccessible.

In the next section, we shall discuss some potential methods which can be used to make an adaptive biometric system along with some preliminary results that was produced using one of the biometric problems.

5.3.1 Ensemble of Classifiers

Since the inception of ensemble-based classification, it has been one of the most studied classification methods [45, 46]. Ensemble-based classifiers have often been used in past for performing incremental/online learning [39–41, 47, 48]. The principle behind the ensemble decision is that the individual predictions combined appropriately should have better overall accuracy, on average, than any individual ensemble member [48]. There are various reasons why an ensemble-based classification is chosen over a single classifier, a few are listed below.

1. No free lunch theorem states that in the absence of prior knowledge about a problem, no one classifier is universally better than any other classifier [49], this also includes random guessing.
2. In case of extremely high-dimensional data, a single classifier's complexity may scale with the dimensionality of the data thus making the generation of a reliable single-classifier infeasible. Instead of a single classifier, generate multiple classifiers on different subsets of features thus reducing the complexity of each classifier trained on the subset.
3. Single classifiers may not work well with data that are too little or too large in size. To work around this problem, ensembles can generate classifiers on multiple bootstrap datasets.
4. Reduces bias towards majority class (class that is well represented by training samples). And generating single strong classifier may be infeasible due to computational costs.

The success of ensemble learning algorithms is believed to depend both on the accuracy and on the diversity among the base learners [50] and some empirical studies revealed that there is a positive correlation between accuracy of the ensemble and diversity among its members [51, 52]. Breiman [53] also shows that random forests with lower generalization error have lower correlation among base learners and higher base learners' strength. Besides, he derives an upper bound for the generalization error of random forests, which depends on both correlation and strength of the base learners.

Literature suggests that there is a tradeoff between base learner's accuracy and diversity, meaning that lower accuracy may indicate higher diversity. However, study in [54] shows that relationship between accuracy, and diversity is not straightforward [55] and lower accuracy may not essentially mean higher diversity. A recent study in [56, 57] discuss that when, how and why ensembles of learning machines can help to handle concept drift in online learning, through a diversity study in the presence of concept drift. This paper presents an analysis of low and high-diversity ensembles combined with different strategies to deal with concept drift and proposes a new approach "Diversity for Dealing with Drifts" (DDD) to handle drifts. DDD maintains [58] ensembles with different diversity levels, exploiting the advantages of diversity to handle drifts and using information from the old concept to aid the learning of the new concept. The authors claim that DDD is accurate both in the presence and in the absence of drifts.

In a recent study, in [59] has reiterated the efficacy of ensemble-based learning to create an adaptive/online learning system for handwritten character recognition. We have used that method to make an adaptive biometric system which learns using one sample at a time. The system as presented in [59] is briefly described below.

The block diagram shown in Fig. 5.1 shows the overall picture of the online learning method proposed in [59]. The method proposed in the paper is to conduct both online supervised as well as online semi-supervised learning. In general, to conduct semi-supervised learning, abundant unlabelled data is required but unfortunately, we have very limited number of samples/class in our biometric learning problem. Therefore, we only conduct online supervised learning but with availability of more data in future, we intend to apply semi-supervised online learning as well. Technical details are given below.

Let us introduce some notation to describe the data. Training dataset $\chi = (X^1, Y^1)(Y^N, Y^N)$, where $X^i = x_1^i x_D^i$, $X^i \in R^D$ are the samples in a D dimensional feature space, and $Y \in 1, K$ are the corresponding labels for a K-class classification problem.

Using Bayes rule and conditional independence among the feature given the class label (assumption used to formulate Naive Bayes classifier), we can write the posterior probability as:

$$P(Y = y_k | x_1 x_D) = \frac{P(Y = y_k) \prod_i P(x_i | Y = y_k)}{\sum_j P(Y = y_j) P(x_1 .... x_D | Y = y_j)}. \tag{5.1}$$

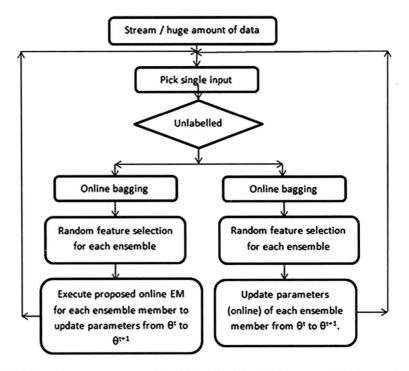

Fig. 5.1 Block diagram representing the batch learning (MLE) of Augmented Naive Bayes network

So to train our classifier we fit a Gaussian $N(x_i; \widehat{\mu_{ik}}, \widehat{\sigma_{ik}^2})$ to each $P(x_i|Y = y_k)$, and we estimate mean and variance for the same using the training data. We perform maximum-likelihood estimation (MLE) to find the mean $\widehat{\mu_{ik}}$ and variance $\widehat{\sigma_{ik}^2}$ of $P(x_i|Y = y_k)$ for each feature x_i, which is just equal to sample mean and sample variance respectively. The classification rule for a new sample $X^{new} = \langle x_1..........x_D \rangle$ can be written as

$$Y^{new} = argmax_{y_k} P(Y = y_k) \prod_i P(x_i|Y = y_k), \quad (5.2)$$

$$Y^{new} = argmax_{y_k} \pi_k \prod_i N(x_i^{new}; \widehat{\mu_{ik}}, \widehat{\sigma_{ik}^2}). \quad (5.3)$$

In order to make an ensemble of B classifiers, we repeat the following steps B times: randomly select F features from the pool of D features. Estimate the learning parameters for the classifier.

A test sample will be classified by each classifier in the ensemble, and the class which gets the majority vote by the ensemble will get assigned to the test sample. In the above explained way, we train an ensemble of B classifiers as an initialization step, by just 1 samples/class. Beyond this with time, as we get more samples of a class, we can update its respective learning parameter as shown below in an online (on the fly) manner using 1 sample at a time. This enables the system to adapt to the changes in underlying distribution of input samples. Initialization (by just one labelled training sample):

$$\widehat{\mu_{ik}} = x_{ik}^1; \widehat{\sigma_{ik}^2} = \sigma_0; \pi_k = \frac{1}{(no. \; of \; classes)}; c_k = 1; \alpha = \alpha_0, \qquad (5.4)$$

where c_k = no. of samples used so far for training,
α = decides the length of memory of the classifier ($\alpha < 1$)
Repeat steps below for all the incoming labelled training samples for any class k:

$$c_k = c_k + 1; \eta_k = \frac{(1 - \alpha)}{c_k} + \alpha, \qquad (5.5)$$

where η_k is learning rate for class k

$$\mu_{ik}(t) = (1 - \eta_k)\mu_{ik}(t - 1) + \eta_k x_i^j \delta(Y^j = y_k) \qquad (5.6)$$

$$\sigma_{ik}^2(t) = (1 - \eta_k)\sigma_{ik}^2(t - 1) + \eta_k (x_i^j - \mu_{ik}(t))^2 \delta(Y^j = y_k). \qquad (5.7)$$

In the same paper [59], another method has been suggested to conduct online learning in semi-supervised manner. The only difference between supervised online learning and semi-supervised online learning lies in the definition of learning rate. Learning rate definition for semi-supervised learning is:

$$\eta_k = q_k \left(\frac{(1 - \alpha)}{c_k} + \alpha\right)\lambda; (\alpha < 1), \qquad (5.8)$$

where λ = weight factor applied to moderate the contribution of unlabelled data in the parameter estimation step.

If an incoming new sample is unlabelled, then the trained classifier is used to produce the posterior $q_k = P(Y^j = y_k | X)$ corresponding to all k (class). This posterior will be used to calculate the learning rate corresponding to all the classes, and subsequently this learning rate will be used to update learning parameters of all the classes as is done in case of supervised online learning.

In [59], the authors have upgraded the Naive Bayes network structure and have suggested method to conduct an online learning for that upgraded network. The network structure was upgraded to get rid of the Naive Bayes unrealistic assumption of conditional independence between different features given the class label. The structure was upgraded with a restriction that all the features will have at most two parents (earlier each had just one) Fig. 5.1. The improved structure was proved to be working much better in the concerned application. The technical detail is briefly explained below (Fig. 5.2).

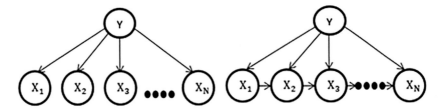

Fig. 5.2 *Left* Bayes net for Naive Bayes. *Right* Bayes net for Augmented Naive Bayes

$P(X_i|X_{i-1}, Y)$ can be parameterized by the following Gaussian distribution: $P(X_i|X_{i-1}, Y) = N(\beta_0 + \beta_1 X_{i-1}, \sigma^2)$.

The author in [59] performs MLE estimation to evaluate the learning parameters β_0, β_1 and σ^2 for supervised learning.

$$\beta_1 = \frac{E_D[X_i X_{i-1}] - E_D[X_i]E_D[X_{i-1}]}{E_D[X_{i-1}X_{i-1}] - (E_D[X_{i-1}])^2} = \Sigma^{-1}_{X_{i-1}X_{i-1}} \Sigma_{X_i X_{i-1}}, \tag{5.9}$$

$$\beta_0 = \mu_{X_i} - \Sigma^{-1}_{X_{i-1}X_{i-1}} \Sigma_{X_i X_{i-1}} \mu_{X_{i-1}}, \tag{5.10}$$

$$\sigma^2 = \Sigma_{X_i X_i} - \Sigma_{X_i X_{i-1}} \Sigma^{-1}_{X_{i-1}X_{i-1}} \Sigma_{X_i X_{i-1}}. \tag{5.11}$$

The authors further propose to update $E_D[X_i]$, $E_D[X_{i-1}]$, $E_D[X_{i-1}X_{i-1}]$, $E_D[X_i X_{i-1}]$ in an online manner in order to update β_0, β_1 and σ^2 because they are the building block which is obvious from their definition. Hence, it can be written as:

$$E_D[X_i](t) = (1 - \eta_k)E_D[X_i](t - 1) + \eta_k X_i, \tag{5.12}$$

$$E_D[X_{i-1}](t) = (1 - \eta_k)E_D[X_{i-1}](t - 1) + \eta_k X_{i-1}, \tag{5.13}$$

$$\sigma^2_{X_{i-1}}(t) = (1 - \eta_k)\sigma^2_{X_{i-1}}(t - 1) + \eta_k(X_{i-1} - E_D[X_{i-1}](t))^2. \tag{5.14}$$

Covariance between two RVs, A and B is given by:

$$\sigma_{(A,B)} = E[AB] - E[A]E[B]. \tag{5.15}$$

Therefore: $E_D[X_{i-1}X_{i-1}](t) = \sigma^2_{X_{i-1}}(t) + E_D[X_{i-1}](t)^2$
Similarly:

$$\sigma_{X_i X_{i-1}}(t) = (1 - \eta_k)\sigma_{X_i X_{i-1}}(t - 1) + \eta_k(X_i - E_D[X_i](t))(X_{i-1} - E_D[X_{i-1}](t)). \tag{5.16}$$

Using the above equations:

$$E_D[X_i X_{i-1}](t) = \sigma_{X_i X_{i-1}}(t) + E_D[X_i](t)E_D[X_{i-1}](t). \tag{5.17}$$

In the equations above, η_k refers to learning rate and its definition is same as in case of Naive Bayes online learning case. And along similar lines, online semi-supervised learning was proposed with a changed definition of learning rate. It has been shown that the upgraded network performs much more accurately as it captures the relationship between different features and accordingly learns. Details can be found in [60].

5.3.2 Incremental/Adaptive Support Vector Machines (SVMs)

SVM is based on a kernel method; however, unlike suboptimal kernel methods, as in the case of a kernel method based on clustering, kernel methods for SVMs are optimal, with the optimality being rooted in convex optimization. Realizing the theoretical strength of SVMs, researchers have developed incremental versions of them. And considering the fact that incremental SVMs have never been explored in the biometric domain, it becomes imperative to discuss SVMs and their incremental versions in the context of adaptive biometrics.

Classification and regression methods based on SVMs [61, 62] are very powerful, which generalize well even in case of very sparse and high-dimensional data. SVM is based on Vapnik's structural risk minimization induction principle which carries out searching over hypothesis classes of varying capacity with best generalization performance.

A two-class classifier based on SVM can be represented as: $f(X) = w.\psi(X) + b$ are learned from the data $\chi = \{(X^1, Y^1)...(Y^N, Y^N)\}$, where $X^i = x_1^i...x_D^i$, $X^i \in R^D$ and $Y^i \in \{-1, 1\}$ by minimizing

$$\min_{w,b,\xi} \frac{1}{2}||w||^2 + C \sum_{i=1}^{N} \xi_i^p. \tag{5.18}$$

For $p \in 1, 2$ subject to the constraints (soft margin)

$$Y^i(w.\psi(X^i) + b) \geq 1 - \xi_i^p, \xi_i^p > 0 \forall i \in (1, ..., N). \tag{5.19}$$

A set of slack variables are introduced for the system to allow few samples to be on the wrong side of the margin (to handle overlapping class distributions) but impose penalty of $\xi_i^p = |Y^i - f(X^i)|^p$ over the objective cost for the samples to be on the wrong side of the margin boundary. Value of $\xi_i^p = 0$ for being on the correct side of the margin boundary. $p = 1$ is what generally preferred in practice because of the robustness to outliers that hinge loss offers as compared to the quadratic loss which corresponds to $p = 2$. The goal is to minimize above the objective function while softly penalizing the points that lie on the wrong side of the margin boundary. Parameter $C > 0$ controls the tradeoff between the slack variable penalty and the margin. Above minimization can be done using quadratic programming but to simplify and take advantage of the kernel trick the above minimization problem is expressed in its dual form

$$\min_{0 \leq \alpha^i \leq C} W = \frac{1}{2} \sum_{i,j=1}^{N} \alpha^i Q^{ij} \alpha^j - \sum_{i=1}^{N} \alpha^i + b \sum_{i=1}^{N} Y^i \alpha^i. \qquad (5.20)$$

With the Lagrange multiplier (and offset) b and $Q^{ij} = Y^i Y^j \psi(X^i).\psi(X^j)$. The above dual form of the original minimization problem must satisfy the famous Karush-Kuhn-Tucker (KKT) condition. The KKT conditions generally involves the primal constrains, dual constrains, and complementary slackness. Therefore, the above dual form along with the KKT condition gives rise to a linearly constrained quadratic programming problem, and there are standard solvers available to solve them. The resulting dual form of the SVM is then $f(X) = \sum_{i=1}^{N} y^i \alpha^i \psi(X^i) \psi(X) + b$. Point to be noted here is that the transformed sample now only appear in dot product. Therefore, one can employ a positive definite kernel function to implicitly project the input samples into some high-dimensional (which can be infinite) space and calculate the dot product to perform classification or regression in that space.

In the mid-90's, support vector machines (SVMs) emerged and subsequently researcher's interest in its online version arose. Early work on this subject by [63] suggests that for each new batch of data, a support vector machine is trained on the new data and the support vectors from the previous learning step. And the logic behind this approach is that the decision function of an SVM depends only on its support vectors, i.e. training an SVM on the support vectors alone results in the same decision function as training on the whole dataset. Because of this, one can expect to get an incremental result that is equal to the non-incremental result, if the last training set contains all examples that are support vectors in the non-incremental case. However, the shortcoming of this approach is that as there are typically only very few support vectors, their influence on the decision function in the next incremental learning step may be very small if the new data is distributed differently.

Note: support vectors are a sufficient description of the decision boundary between the examples, but not of the examples themselves.

The above problem was addressed in [64] by making a clever change in the objective function to be optimized, and i.e. by making the error on old support vectors (which represent the old learning set) more costly than an error on a new example. Details can be found on the concerned paper. At the same time, [53] exploits the locality of the RBF kernel to build online SVM. The authors do not use all the previous support vectors (as done in [63, 64]), instead it only uses the support candidates in the neighbourhood of the new incoming sample. Although deciding the neighbourhood is critical, the method would be fast. However, above three approaches and methods proposed by [65]; provide only approximate solution and may require many passes through the dataset to reach a reasonable level of convergence.

An exact solution to the problem of online SVM learning has been found by [66]. Their incremental algorithm updates an optimal solution of an SVM training problem after one training example is added (or removed). In this, the authors construct the solution recursively one point at a time such that the KKT condition is satisfied over all the data already seen as well as the new incoming samples.

The first-order condition on W reduces to KKT condition:

$$g^i = \frac{\partial W}{\partial \alpha^i} = \sum_{j=1}^{N} Q^{ij}\alpha^j + Y^i b - 1 = Y^i f(X^i) - 1 \qquad (5.21)$$

$$g^i = \begin{cases} \geq 0 & \alpha^i = 0 \\ = 0 & 0 < \alpha^i < C \;] \\ \leq 0 & \alpha^i = C \end{cases} \qquad (5.22)$$

$$\frac{\partial W}{\partial b} = \sum_{j=1}^{N} y^j \alpha^j = 0. \qquad (5.23)$$

This partitions the data into three categories:

(A) $x^i \in S \subset \chi$ where S is a set of margin support vectors, strictly on the margin (i.e. $Y^i f(X^i) = 1$).
(B) $x^i \in O \subset \chi$, where O is the set of other vectors for which $Y^i f(X^i) > 1$ i.e. the sample is on the correct side of the margin boundary (correctly classified).
(C) $x^i \in E \subset \chi$ where E is a set of error vectors $Y^i f(X^i) < 1$, sample is on the wrong side of the margin boundary but not necessarily mis-classified.

The set $R = \{O \cup E\}$ is a set of reserve vectors. Lower case letters s, e, o and r will be used to refer to such kind of partitions.

By writing the KKT conditions before and after an update $\Delta\alpha$, we obtain the following conditions that must be satisfied after an update [66]:

$$\begin{bmatrix} \Delta g^c \\ \Delta g^s \\ \Delta g^r \\ 0 \end{bmatrix} = \begin{bmatrix} Y^c & Q^{cs} \\ Y^s & Q^{ss} \\ Y^r & Q^{rs} \\ 0 & Y^{sT} \end{bmatrix} \begin{bmatrix} \Delta b \\ \Delta \alpha^s \end{bmatrix} + \Delta\alpha^c \begin{bmatrix} Q^{csT} \\ Q^{ssT} \\ Q^{rsT} \\ Y^c \end{bmatrix} \qquad (5.24)$$

It is easy to see that $\Delta\alpha^c$ is in equilibrium with $\Delta\alpha^s$ and Δb in order for the above condition to hold. Considering the fact that $\Delta g^s = 0$, from line 2 and 4 of the above equation we can write:

$$\begin{bmatrix} 0 \\ 0 \end{bmatrix} = \begin{bmatrix} 0 & Y^{sT} \\ Y^s & Y^{ss} \end{bmatrix} \Delta s + \begin{bmatrix} Y^c \\ Q^{csT} \end{bmatrix} ; where \; \Delta s = \begin{bmatrix} \Delta b \\ \Delta \alpha^s \end{bmatrix} \qquad (5.25)$$

Above linear equation can be solved for Δs

$$\Delta s = \beta \Delta\alpha^c \qquad (5.26)$$

where

$$\beta = - \begin{bmatrix} 0 & Y^{sT} \\ 0 & Q^{ss} \end{bmatrix}^{-1} \begin{bmatrix} Y^c \\ Q^{csT} \end{bmatrix} \qquad (5.27)$$

is the gradient of the manifold of optimal solutions parameterized by α^c Similarly from line 1 and 3

$$\begin{bmatrix} \Delta g^c \\ \Delta g^r \end{bmatrix} = \gamma \Delta \alpha^c \qquad (5.28)$$

where

$$\gamma = \begin{bmatrix} \gamma^c & Q^{cs} \\ \gamma^c & Q^{rs} \end{bmatrix} \beta + \begin{bmatrix} Q^{ccT} \\ Q^{csT} \end{bmatrix} \qquad (5.29)$$

is the gradient of the manifold of gradient g^r at an optimal solution parameterized by α^c. These refined set of equations shows that update process is controlled by very simple sensitivity relation: $\Delta s = \beta \Delta \alpha^c$ and $[\Delta g^c \Delta g^r]^T = \gamma \Delta \alpha^c$, where β is the sensitivity of Δs with respect to $\Delta \alpha^c$ and γ is the sensitivity of Δg^c and Δg^r with respect to $\Delta \alpha^c$.

At this stage to carry out the parameter update process, the key is to find out the largest possible increment of $\Delta \alpha^c$ and subsequently Δs and Δg is updated. Authors of [66] have very exhaustively addressed all the cases by which one can determine the largest value of $\Delta \alpha^c$. And once the step is determined, one can follow the steps of algorithm 1 given in [67] to carry out online learning.

Finding an absence of a well-accepted implementation of the work by [66, 67] proposed a new design of storage and numerical operation which speeds up the incremental SVM training by a factor of 5–20. On the similar line, [68] have applied the accumulated knowledge of optimization to the computational problem presented by the SVM to propose a very efficient way of training SVM in incremental fashion.

5.3.3 Incremental/Adaptive Neural Network

Neural networks being one of the oldest methods of machine learning, it is very obvious that umpteen amount of work have been done on that domain and many of them deals with incremental and adaptive learning. Here we list few briefly which can be useful to create an adaptive biometric system.

Fuzzy ARTMAP: This is a neural network-based structure, and it is one of the earliest methods used in incremental learning. The fuzzy ARTMAP has two fuzzy ART modules that are linked via an inter-art module known as "map field" The map field is used to form predictive categories for learning class association. Fuzzy ARTMAP will generate new decision clusters in response to new input patterns that are sufficiently different from previously seen instances. The'sufficiently different' patterns are controlled using a free parameter of ARTMAP known as the vigilance parameter. ARTMAP is sensitive to the vigilance parameter especially in the presence of significant noise in the training data. Using stability and match tracking, fuzzy ARTMAP automatically constructs as many categories as are needed to learn any static training set to 100 %. Thus, fuzzy ARTMAP may over-fit, leading to poor generalization.

Learn++: This is one of the most notable families of incremental learning algorithm, which was first introduced by Polikar et al. [69] and later upgraded by few other authors for e.g. [39]. It creates multiple classifiers to each data chunk presented to the system. Inspired by AdaBoost [57, 70] for each chunk, the training set for each base learner is created by sampling examples according to a distribution of probability. Like AdaBoost, Learn++ maintains a distribution of instance weights; however, Learn++ does not update the weights in the same manner as performed with AdaBoost. In AdaBoost, the distribution of probability is built to give higher priority to instances mis-classified by the last previously created classifier, whereas Learn++ uses the ensemble decision, rather than the decision of the latest classifier. When a new dataset arrives, the distribution is re-initialized by evaluating the entire ensemble and initializing the distribution. Pros: Learn++ does not have to access the previous data chunks, and it demonstrates considerable improvement at generalization when compared with fuzzy ARTMAP on common databases. Cons: problem is that a new set of classifiers is created for each new data chunk. So, the ensemble size can become extremely large considering lifelong learning.

Self-Organized Incremental Neural Network (SOINN): This is an unsupervised incremental learning method which was proposed in [68] for topology learning and classification to handle noisy unlabelled data. This method is essentially a combination of self-organizing map [71] and competitive Hebbian learning [72] which can be used to learn the topology of the input data stream. The proposed algorithm makes a two-layered neural network Fig. 5.3. The first layer which represents a reasonable topological structure of unlabelled data gives a reasonable number of clusters and gives a typical prototype pattern of every cluster. Prior knowledge regarding the number of classes or codebook is not required.

The first layer learns the density distribution of the input pattern. Subsequently, output of the first layer serves as the input of the second layer, where the different clusters are separated by detecting the low-density overlap area. The method uses similarity threshold and a locally accumulated error-based insertion criterion to grow the system incrementally and accommodate the input patterns of online non-stationary data distribution. It also uses and online criterion to delete nodes from the low-probability regions, and this enables the system to separate the cluster and simultaneously eliminate the noise noisy samples from the input data. Authors use "error radius" as the utility parameters to control the growth of number of nodes in the network and check successful node insertion. Although the method has been successfully applied on some real-time problems, it has several limitations for e.g. (a) in case of high-density overlap, it is difficult for the method to separate the clusters, and (b) several important parameters value need to be decided by the users which increase the chances that the system getting used sub-optimally. On the similar lines authors have further modified the SOINN to make a semi-supervised incremental

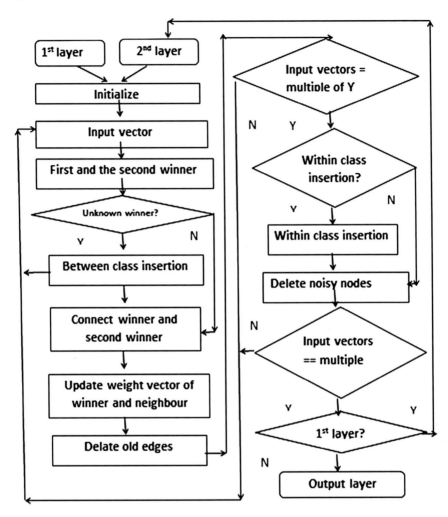

Fig. 5.3 Flowchart of SOINN, shows the basic overflow of the proposed algorithm

active learning system [73] which is very promising and claims have been made that this method can be very useful to create a never ending learning system.

5.4 Experimental Details and Results

In order to evaluate the effectiveness of the discussed incremental learning algorithms, we have used one of the discussed methods to develop an adaptive sclera biometric system. Fuzzy C-means clustering was used to segment the sclera [74–76].

Since the vessel patterns in the sclera are not prominent, the image was enhanced to make it visible. A new preprocessing approach for vein highlighting is proposed using adaptive histogram equalization [77]. And histogram equalization was performed with a window size of 42×42 (the window value was selected by analysis, window value that produces the best result was used for experimentation). Since the sclera vessel patterns are most prominent in the green channel, it was used for preprocessing to make the vessel structure more prominent

Furthermore, a bank of filter based on Discrete Meyer wavelet [78] was used to enhance the vessel patterns. Low-pass reconstruction of the above-mentioned filter was used to enhance the image.

We use sclera feature extraction based on the dense local directional pattern (D-LDP) [79] which is an extended version of LDP proposed in [80]. For extracting the patch/dense descriptors, each image was divided into a regular dense grid of three different special pyramidal planes. A higher-order LDP with 4 orientations was used here for the feature extraction.

A histogram of bin size 512, for each of these patches are calculated in multi-scales of 1, 2 and 3. Finally the histogram of each patch and each scale are calculated and concatenated column wise to get the final descriptor.

The irises were segmented along the radius by calculating the centre and the radius by the integro-differential [81] operator and further enhanced using an adaptive histogram equalization technique. The red channel of the colour image was used for iris image enhancement. Image level fusion was performed using iris and sclera, and subsequently, the patterns were classified using the developed features.

In order to perform the experiment, UBIRIS version 1 [82] dataset was used in the experiment. This database consists of 1877 RGB images taken in two different sessions (1205 images in session 1 and 672 images in session 2), from 241 identities and images are represented in RGB colour space. The database contains blurred images and images with blinking eyes. Both high-resolution images (800×600) and low-resolution images (200×150) are present in the database. For each individual, 10 eye images were present. All the images are in JPEG format.

The dataset consists of different quality of images with respect to sclera region visibility. Some of the images are not occluded and have sclera regions visible which are of good quality; some of them are of medium quality and the third type was of poor quality with respect to sclera region visibility as shown in Fig. 5.4.

First session images have minimum noise in the form of reflections, luminosity and contrast, as it uses a standard image capturing framework inside a closed room with controlled artificial lights. In the second session, pictures were captured in more natural conditions, for example, it uses both natural and artificial light, there by bringing in the changing image capturing settings into play. This difference in the capturing environment produces heterogeneous images with respect to reflections, contrast, luminosity and focus. Images collected at second session were captured by a vision system with or without minimal active participation from the subjects, adding several noise problems. There was a gap of two weeks between the two session.

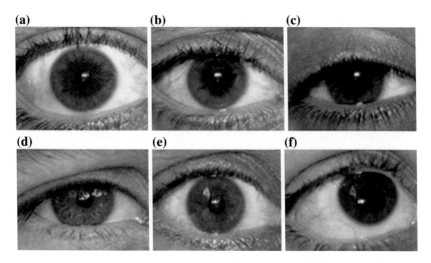

Fig. 5.4 Different quality of eye images used in experiment (**a**) is the type of best quality images of Session 1, **b** is the type of medium quality of Session 1 (**c**) is the type of Poor quality of Session 1, **d** is the type of below average quality image of Session 2, section is the type of average quality of Session 2 (**f**) is the type of best quality image in Session 2

These changes in environmental condition in the form of different gaze angle of the eye, data accruing techniques and time-span gap were utilized to investigate the adaptability of the sclera texture for biometric identification.

It was observed that the number of participants in the two sessions were not the same which produced a very uneven number of samples corresponding to different individuals. Even the number of the population was different for the two sessions. The first session consisted of 241 users and in the second session there were 135 users.

We have used 50 out of 241 total classes present because many classes did not have data from both sessions which makes the number of samples too few to apply the learning algorithm over them. We did not use a few classes because the iris and sclera region of the participants were too occluded to be used for learning. Few examples of such images are given below in Fig. 5.5.

For our experiment as a learner, we used incremental Naive Bayes classifier as proposed in [59]. The number of features used for each class was 30480. The total number of classes used was 50. For each feature, we fitted a single Gaussian of the form $P(x_i|Y = y_k)$ corresponding to each class. Where i and k refers to ith feature and kth class, respectively. The classifier is initially trained offline with few numbers of samples using maximum-likelihood estimation. Subsequently, we updated the learning parameters of each Gaussian corresponding to each class in the online fashion as discussed in Sect. 5.3.1 and [59]. The value of constant α was empirically decided to be 0.55 for the experiments. We did not create ensemble of classifiers and semi-supervised learning as we did not have enough data per class to create different

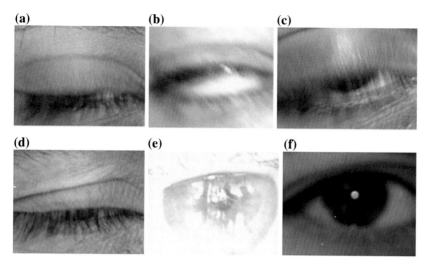

Fig. 5.5 Example of closed and blurred eyes. **a–c** are of session 1 and **d–f** are of session 2

classifiers. But creating an ensemble using a randomization technique is a powerful technique to create a boost in accuracy of the classifier. We shall try this once we have more data.

Few parameters used in the learning method employed are as follows:

(a) c_k which keeps track of the number of samples that have been used for each class for online training.

(b) α it is a constant which decides the length of the memory of the classifier, it influences the value of η_k which decides how much weight must be given (in general and when the convergence (when $c_k \rightarrow \infty$) is achieved) to the new incoming samples in the learning parameter estimation step.

(c) η_k is the learning rate parameter whose value depends on the value of count of samples and the value of α and the value of c_k. The detailed discussion over the role of these parameters in learning process is given in [59].

The results table shows the accuracy of the applied method under different settings of the experiments. Considering the fact that Naive Bayes is a classifier which is based upon a very strong assumption of conditional independence among the features given the class, it is performing reasonably well. Therefore, it can be assumed that the stronger version of incremental learning algorithms could perform even better. This experimental process shows that larger accuracies can be achieved using incremental learning in the biometric domain by more experimentation with other adaptive learning techniques available in the machine learning literature (Table 5.1).

Table 5.1 Results show that adaptive/online Naive Bayes classifier is more accurate

Classifier name	Training/Testing	Accuracy (%)
Naive Bayes (NB)	7 offline/3	60
Online/adaptive NB	5 offline + 2 online/3	72
Online/adaptive NB	5 offline + 3 online/2	86
Online/adaptive NB	5 offline + 4 online/1	88

Adaptive classifiers are initially trained with a few samples in batch/offline mode and later learn in an online/adaptive manner i.e. using one sample at a time. For example: row 2:- classifier was initially trained with 5 samples per class, and sub-sequently, it was adaptively trained with 2 samples per class and later tested with 3 samples per class

5.5 Conclusions

In this chapter, we discussed many potential existing adaptive learning methods, which can be applied in the biometric domain to create a robust adaptive biometric system. To demonstrate this, we applied one existing learning method to create an adaptive multi-modal ocular approach using iris and sclera. It is evident from the experiments that the adaptive/incremental system applied outperforms the base classifier performance. It is also evident from the results that when the numbers of samples are increased, the adaptability also improves. Due to lack of availability of large number of samples, other theories like online semi-supervised learning methods were not tested. In future, we plan to explore all the promising adaptive techniques in the biometric domain and build a reliable never ending learning biometric system. In that regard, we are in the process of collecting more data so that more sophisticated learning methods based on incremental semi-supervised learning can be applied to gain the leverage from the availability of large data.

References

1. Rattani, A., Freni, B., Marcialis, G.L., Roli, F.: Template update methods in adaptive biometric systems: a critical review. In: Proceedings of Third International Conference Biometrics, Sardinia, Alghero, pp. 847–856 (2009)
2. Rattani, A., Marcialis, G.L., Granger, E., Roli, F.: A Dual-staged Classification-selection approach for automated update of biometric templates. In: International Conference on Pattern Recognition (ICPR), pp. 2972–2975. Japan (2012)
3. Rattani, A.: Adaptive biometric system based on template update procedures. Ph.D. thesis, University of Cagliari, Italy (2010)
4. Rattani, A., Freni, B., Marcialis, G.L., Roli, F.: Template update methods in adaptive biometric systems: a critical review. In: Proceedings of Third International Conference Biometrics, pp. 847–856. Sardinia, Alghero (2009)
5. Poh, N., Kittler, J., Marcel, S., Matrouf, D., Bonastre, J.F.: Model and score adaptation for biometric systems: coping with device interoperability and changing acquisition conditions. In: Proceedings of the International Confernece on Pattern Recognition, pp. 1229–1232. Istambul, Turkey (2010)

6. Uludag, U., Ross, A., Jain, A.: Biometric template selection and update: a case study in finger-prints. Pattern Recogn. **37**(7), 1533–1542 (2004)
7. Oh, K., Toh K.: Extracting sclera features for cancelable identity verification. In: 5th IAPR International Conference on Biometric, pp. 245–250 (2012)
8. Jiang, X., Ser, W.: Online fingerprint template improvement. IEEE Trans. Pattern Anal. Mach. Intell. **24**(8), 1121–1126 (2008)
9. Ryu, C., Hakil, K., Jain, A.K.: Template adaptation based finger print verification. In: Proceedings of 18th International Conference on Pattern Recognition, pp. 582–585. Hong Kong (2006)
10. Liu, X., Chen, T., Thornton, S.M.: Eigenspace updating for non stationary process and its application to face recognition. Pattern Recognit. **36**(9), 1945–1959 (2003)
11. Roli, F, Didaci, L., Marcialis, G.L.: Template co-update in multimodal biometric systems. In: Proceedings of IEEE/IAPR International Conference on Biometrics, pp. 1194–1202. Seoul, Korea (2007)
12. Roli, F., Marcialis, G.L.: Semi-supervised PCA-based face recognition using self-training. In: Proceedings of the Joint IAPR International Workshop on S+SSPR06, pp. 560–568. Hong Kong, China (2006)
13. Rattani, A., Marcialis, G.L., Roli, F.: Biometric template update usingthe graph mincut: a case study in face verification. In: Proceedings of Sixth IEEE Biometric Symposium, pp. 23–28. Tampa, USA (2008)
14. Poh, N., Rattani, A., Roli, F.: Critical analysis of adaptive biometric systems. IET Biometrics **1**(4), 179–187 (2012)
15. Das, A.: Face recognition in reduced eigen plane. In: International Conference on Communications, Devices and Intelligent Systems, pp. 620–623 (2012)
16. Das, A., Pal, U., Ballester, M.F., Blumenstein, M.: A new wrist vein biometric system. In: Computational Intelligence in Biometrics and Identity Management (CIBIM), pp. 68–75 (2014)
17. Das, A., Parekh, R.: Iris recognition using a scalar based template in eigenspace. Int. J. Comput. Sci. Telecommun. **3**(5), 74–79 (2012)
18. Das, A., Parekh, R.: Iris recognition in 2D eigen-space. Int. J. Comput. Appl. **52**(19), 1–6 (2012)
19. Derakhshani, R., Ross, A., Crihalmeanu, S.: A new biometric modality based on conjunctival vasculature. In: Proceedings of Artificial Neural Networks in Engineering, pp. 1–8 (2006)
20. Khosravi, M.H., Safabakhsh, R.: Human eye sclera detection and tracking using a modified time-adaptive self-organizing map. Pattern Recogn. **41**, 2571–2593 (2008)
21. Zhou, Z., Du, Y., Thomas, N.L., Delp, E.J.: A new biometric sclera recognition. IEEE Trans. Syst. Man Cybern. -PART A: Syst. Hum. **42**(3), 571–583 (2012)
22. Zhou, Z., Du, Y., Thomas, N.L., Delp, E.J.: Quality fusion based multi-modal eye recognition. In: IEEE International Conference on Systems, Man, and Cybernetics, pp. 1297–1302 (2012)
23. Das, A., Pal, U., Ballester, M.F., Blumenstein, M.: Fuzzy logic based sclera recognition. In: FUZZ-IEEE, pp. 561–568 (2014)
24. Tankasala, S.P., Doynov, P., Derakhshani, R.R., Ross, A., Crihalmeanu, S.: Biometric recognition of conjunctival vasculature using GLCM features. In: International Conference on Image Information Processing, pp. 1–6 (2011)
25. Zhou, Z., Du, Y., Thomas, N.L., Delp, E.J.: Multi angled sclera recognition. In: IEEE Workshop on Computational Intelligence in Biometrics and Identity Management, pp. 103–108 (2011)
26. Crihalmeanu, S., Ross, A.: Multispectral sclera patterns for ocular biometric recognition. Pattern Recogn. Lett. **33**, 1860–1869 (2012)
27. Das, A., Pal, U., Ballester, M.F.A., Blumenstein, M.: A new method for sclera vessel recognition using OLBP. In: Chinese Conference on Biometric Recognition, LNCS, vol. 8232, pp. 370–377 (2013)
28. Ferrer, M.A., Morales, A., Das, A., Blumenstein, M., Pal, U.: Model based sclera vessels segmentation with SIFT recognition and its combination with Iris. In: Spanish biometric consodium, VII Jornadas de Reconocimiento Bio-metrico de Personas, pp. 68–76 (2013)

29. Das, A., Pal, U., Ballester, M.F., Blumenstein, M.: Sclera recognition using D-SIFT. In: 13th International Conference on Intelligent Systems Design and Applications, pp. 74–79 (2013)
30. Lazebnik, S., Schmid, C., Ponce, J.: Beyond bags of features: spatial pyramid matching for recognizing natural scene categories. Proc. Comput. Vis. Pattern Recogn. **2**, 2169–2178 (2006)
31. Das, A., Pal, U., Ballester, M.F., Blumenstein, M.: A new efficient and adaptive sclera recognition system. In: Computational Intelligence in Biometrics and Identity Management (CIBIM), pp. 1–8 (2014)
32. Das, A., Pal, U., Ballester, M.F., Blumenstein, M.: Multi angle based lively sclera biometrics at a distance. In: Computational Intelligence in Biometrics and Identity Management (CIBIM), pp. 22–29 (2014)
33. Zhou, Z., Du, Y., Thomas, N.L., Delp, E.J.: Multimodal eye recognition. Proc. Int. Soc. Optical Eng. **7708**(770806), 1–10 (2010)
34. Gottemukkula, V., Saripalle, S.K., Tankasala, S.P., Derakhshani, R., Pasula, R., Ross, A.: Fusing iris and conjunctival vasculature: ocular biometrics in the visible spectrum. In: IEEE Conference on Technologies for Homeland Security, pp. 150–155 (2012)
35. Das, A., Pal, U., Blumenstein, M., Ballester, M.F.: Sclera recognition—a survey. Adv. Comput. Vis. Pattern Recogn. 917–921 (2013)
36. Scholz, M., Klinkenberg, R.: Boosting classifiers for drifting concepts. In: IDA—Special Issue on Knowledge Discovery from Data Streams, vol. 11, Issue 1, pp. 3–28 (2007)
37. Rodrigues, P.P., Gama, J.A., Arajo, J.A., Lopes, L.: L2GClust: local-to-global clustering of stream sources. In: Proceedings of the 2011 ACM Symposium on Applied Computing, pp. 1006–1011. ACM (2011)
38. Kolter, J.Z., Maloof, M.A.: Dynamic weighted majority: an ensemble method for drifting concepts. J. Mach. Learn. Res. **8**, 2755–2790 (2007)
39. Widmer, G., Kubat, M.: Learning in the presence of concept drift and hidden contexts. Mach. Learn. **23**, 69–101 (1996)
40. Street, N.W., Kim, Y.: A streaming ensemble algorithm (SEA) for large-scale classification. In: KDD '01: Proceedings of the Seventh ACM SIGKDD International Conference on Knowledge Discovery and Data Mining, pp. 377–382. ACM (2001)
41. Klinkenberg, R.: Learning drifting concepts: example selection vs. example weighting. Intell. Data Anal. **8**, 281–300 (2004)
42. Fern, A., Givan, R.: Online ensemble learning: an empirical study. Mach. Learn. **53**, 71–109 (2003)
43. Maloof, M.A., Michalski, R.S.: Incremental learning with partial instance memory. Artif. Intell. **154**(1–2), 95–126 (2004)
44. Shilton, A., Palaniswami, M., Ralph, D., Tsoi, A.C.: Incremental training of support vector machines. IEEE Trans. Neural Networks **16**, 114–131 (2005)
45. Kuncheva, L.I.: Combining Pattern Classifiers: Methods and Algorithms. Wiley (2004)
46. Brown, G.: Ensemble learning. In: Encyclopedia of Machine Learning, pp. 1–9 (2010)
47. Polikar, R.: Bootstrap inspired techniques in computational intelligence. IEEE Signal Process. Mag. **24**, 57–72 (2007)
48. Wolpert, D.H., Macready, W.G.: No free lunch theorems for optimization. IEEE Trans. Evol. Comput. **1**(1), 67–82 (1997)
49. Dietterich, T.G.: Machine learning research: four current directions. Artif. Intell. **18**(4), 97–136 (1997)
50. Dietterich, T.G.: An experimental comparison of three methods for constructing ensembles of decision trees: bagging, boosting, and randomization. Mach. Learn. **40**(2), 139–157 (2000)
51. Kuncheva, I., Whitaker, C.J.: Measures of diversity in classifier ensembles and their relationship with the ensemble accuracy. Mach. Learn. **51**, 181–207 (2003)
52. Breiman, L.: Random forests. Mach. Learn. **45**, 5–32 (2001)
53. Ralaivola, L., d'Alche Buc, F.: Incremental support vector machine learning: a local approach. Lect. Notes Comput. Sci. **2130**, 322–329 (2001)
54. Minku, L.L., White, A., Yao, X.: The impact of diversity on online ensemble learning in the presence of concept drift. IEEE Trans. Knowl. Data Eng. **22**, 730–742 (2010)

55. Tang, E.K., Sunganthan, P.N., Yao, X.: An analysis of diversity measures. Mach. Learn. **65**, 247–271 (2006)
56. Martinetz, T.M., Berkovich, S.G., Schulten, K.J.: Neural-gas network for vector quantization and its application to time-series rrediction. IEEE Trans. Neural Networks **4**(4), 558–569 (1993)
57. Freund, Y., Schapire, R.E.: Experiments with a new boosting algorithm. In: Proceedings of the Thirteenth International Conference on Machine Learning (ICML'96), pp. 148–156. Morgan Kaufmann, Bari, Italy (1996)
58. Minku, L.L., Yao, X.: DDD: a new ensemble approach for dealing with concept drift. IEEE Trans. Knowl. Data Eng. (2010)
59. Kunwar, R., Pal, U., Blumenstein, M.: Semi-supervised online learning of handwritten characters using a bayesian classifier. In: Second IAPR Asian Conference on Pattern Recognition, pp. 717–721. Okinawa, Japan (2013)
60. Kunwar, R., Pal, U., Blumenstein, M.: Semi-supervised online bayesian network learner for handwritten characters recognition. In: Twenty Second International Conference on Pattern Recognition (ICPR), pp. 3104–3109. Stockholm, Sweden (2014)
61. William, H., Saul, A., William, T., Flannery, B.P.: Support Vector Machines. Numerical Recipes: The Art of Scientific Computing, 3rd edn. Cambridge University Press, New York (2007). ISBN 978-0-521-88068-8
62. Vladimir, V., Vapnik, V.: Statistical Learning Thoery. Springer, New York (1998)
63. Syed, N.A., Liu, H., Sung, K.K.: Incremental learning with support vector machines. In: SVM workshop, IJCAI (1999)
64. Ruping, S.: Incremental learning with support vector machines. Technical Report TR-18, Universitat Dortmund, SFB475 (2002)
65. Kivinen, J., Smola, A.J., Williamson, R.C.: Online learning with kernels. In: Diettrich, T.G., Becker, S., Ghahramani, Z. (eds.) Advances in Neural Information Proceedings Systems (NIPS 01), pp. 785–792 (2001)
66. Cauwenberghs, G., Poggio, T.: Incremental and decremental support vector machine learning. In: Leen, T.K., Dietterich, T.G., Tresp, V. (eds.) Advances in Neural Information Processing Systems, vol. 13, pp. 409–415. MIT Press (2001)
67. Laskov, P., Gehl, C., Kruger, S., Muller, K.R.: Incremental support vector learning: analysis, implementation and applications. J. Mach. Learn. Res. 1909–1936 (2006)
68. Shilton, A., Palaniswami, M., Ralph, D., Tsol, A.C.: Incremental training of support vector machines. IEEE Trans. Neural Networks **16**(1), 114–131 (2005)
69. Muhlbaier, M., Topalis, A., Polikar, R.: Learn++. NC: combining ensemble of classifiers combined with dynamically weighted consult-and-vote for efficient incremental learning of new classes. IEEE Trans. Neural Networks **20**(1), 152–168 (2009)
70. Freund, Y., Schapire, R.: A decision-theoretic generalization of on-line learning and an application to boosting. J. Comput. Syst. Sci. **55**(1), 119–139 (1997)
71. Shen, F., Sakurai, K., Kamiya, Y., Hasegawa, O.: An online semi-supervised active learning algorithm with self-organizing incremental neural network. In: International Joint Conference on Neural Network (IJCNN), pp. 1139–1144 (2007)
72. Kohonen, T.: Self-organized formation of topologically correct feature maps. Biol. Cybern **43**(1), 59–69 (1982)
73. Shen, F., Hasegawa, O.: An incremental network for on-line unsupervised classification and topology learning. Neural Networks **19**(1), 90–106 (2006)
74. Dunn, J.C.: A fuzzy relative of the ISODATA process and its use in detect-ing compact well-separated clusters. J. Cybern. **3**, 32–57 (1973)
75. Bezdek, J.C.: Pattern Recognition with Fuzzy Objective Function Algorithms. Plenum Press, New York (1981)
76. Li, B.N., Chui, C.K., Chang, S., Ong, S.H.: Integrating spatial fuzzy clustering with level set methods for automated medical image segmentation. Comput. Biol. Med. **41**(1), 1–10 (2011)
77. Pizer, S.M., Amburn, E.P., Austin, J.D.: Adaptive histogram equalization and its variations. Comput. Vis. Graph. Image Process. **39**, 355–368 (1987)

78. Daubechies, I.: Ten lectures on wavelets. In: CBMS-NSF Conference Series In Applied Mathematics, SIAM Ed, pp. 117–119 (1992)
79. Zhang, B., Gao, Y., Zhao, S., Liu, J.: Local derivative pattern versus local binary pattern: face recognition with high-order local pattern descriptor. IEEE Trans. Image Proces. **19**(2) (2010)
80. Kabir, Md. H., Jabid, T., Chae, O.: A local directional pattern variance (LDPv) based face descriptor for human facial expression recognition. In: Proceedings of the IEEE Advanced Video and Signal Based Surveillance (AVSS), pp. 526–532 (2010)
81. Daugman, J.G.: High confidence visual recognition of persons by a test of statistical independence. IEEE Trans. Pattern Anal. Mach. Intell. **15**(11), 1148–1161 (1993)
82. Proena, H., Alexandre, L.A.: UBIRIS: a noisy iris image database. In: Proceedings of ICIAP: International Conference on Image Analysis and Processing, vol. 1, pp. 970–977 (2005)

Chapter 6
Adaptive Facial Recognition Under Ageing Effect

**Zahid Akhtar, Amr Ahmed, Cigdem Eroglu Erdem
and Gian Luca Foresti**

Abstract Being biological tissue in nature, facial biometric trait undergoes ageing. Previous studies indicate that ageing has profound effects on face biometrics as it causes change in shape and texture. Despite the rising attention to facial ageing, longitudinal study of face recognition remains an under-studied problem in comparison to facial variations due to pose, illumination and expression changes. A commonly adopted solution in the state-of-the-art is the virtual template synthesis for ageing and de-ageing transformations involving complex 3D modelling techniques. However, these schemes are prone to estimation errors in the synthesis. Another promising solution is to continuously adapt the enrolled templates to the temporal variation (ageing) of the input samples based on some learning methodology. Although efficacy of template update procedures has been proven for expression, lightning and pose variations, the use of template update for facial ageing has been mainly overlooked till date. To this aim, the contributions of this chapter are (a) evaluation of *six* baseline facial representations, based on local features, under the *ageing effect*, (b) analysis of the compound effect of ageing with other variates, i.e. race, gender, glasses, facial hair etc., (c) introducing template ageing as a concept drift problem, and (d) investigating the use of template update procedures for temporal variance due to the facial ageing process.

Z. Akhtar (✉) · G.L. Foresti
University of Udine, Udine, Italy
e-mail: zahid.akhtar@uniud.it

G.L. Foresti
e-mail: gianluca.foresti@uniud.it

A. Ahmed
University of Lincoln, Lincoln, UK
e-mail: aahmed@lincoln.ac.uk

C.E. Erdem
Bahcesehir University, Istanbul, Turkey
e-mail: cigdem.eroglu@eng.bahcesehir.edu.tr

© Springer International Publishing Switzerland 2015
A. Rattani et al. (eds.), *Adaptive Biometric Systems*,
Advances in Computer Vision and Pattern Recognition,
DOI 10.1007/978-3-319-24865-3_6

6.1 Introduction

Face plays a vital role in our social interaction, bearing a person's identity. Face recognition (FR) for biometric use has received significant attention in the past several years due to many appealing qualities such as universality, acceptability, easy collectability and wide variety of applications in both law enforcement and non-law enforcement. Face recognition is a non-intrusive technique and can be used with existing image capture devices (web-cams, security cameras, etc.), thus enabling FR to have a covert or surveillance (CCTV) capability. Nowadays, use of the human face as a key to security has been consolidated in various biometric identity management programs, such as USVISIT, UIDAI, National ID cards, consumer ID, etc. Nevertheless, with the rapid gain in the use and popularity of facial biometrics, there also exist a need of more robust, reliable and accurate face recognition system [1–5].

A facial biometric verification system consists of two main processes; enrolment and verification. In enrolment, individual's face samples are captured and preprocessed, and features are then extracted. These extracted features are labelled with user's ID called the "template", representing his identity. Verification mode verifies claimed identity by comparing input sample(s) to the enrolled template(s). The efficiency of the facial biometric systems depends on the representative capability of the enrolled templates. Performance of these systems can be measured in terms of false acceptance rate (FAR) and false rejection rate (FRR).

The compound effect of the inherent scarcity of training samples during the enrolment phase and the presence of substantial sample variations during the operational phase is the major cause of errors in face recognition systems. According to the latest report by Facial Recognition Vendor Test (FRVT) 2013 [6], there has been improvement in the order of magnitude in *controlled environment* since FRVT 2006. However, FR systems still perform poor under uncontrolled environment, namely variations caused in the input data by pose, illumination and expression, etc. It is worth noting that large sample variation is caused by the vulnerable nature of data acquisition process and the external changing acquisition conditions, change of sensor, etc. Apart from the above factors, being biological tissues in nature, face biometric trait can be altered either temporarily or permanently due to ageing, diseases, treatment to diseases, injuries or plastic/cosmetic surgery. Among them face ageing is a progressive accumulation of changes with time, and how fast we age varies from one individual to another. Ageing effects both shape and texture, and is usually contributed by our genes, environmental influences and life style. An important consequence of the ageing process is that enrolled templates become unrepresentative of the input (query) data after a certain time lapse as a result of change in the data distribution of an individual. This establishes its similarity with the concept drift theory [7, 8], dealing with changing concept (models) over time and offering variant update procedures as the solution. Existing studies [9–12] on different databases and for different algorithms report evidences of the performance degradation of face recognition systems as a result of time lapse between the pair of facial images. Nevertheless, facial ageing has not received substantial attention in comparison to other facial variations.

Existing standard age-invariant solution is the virtual biometric synthesis of ageing and de-ageing transformation based on the simulation of the craniofacial morphological changes [13–16], including complex 3D modelling techniques. FGNET [17], MORPH [18] and BROWNS [19] are commonly adopted databases for the evaluation of these methods. However, performance gain obtained due to these methods may be limited due to mainly two reasons. First, ageing is a complex process which differs from person to person and occurs in different manifestation in different age groups. Second, the simulation process is affected by other variations such as facial hairs, glasses, pose, lightning and expression, as present in the databases. Therefore, these factors might cause the age transformation procedure to be prone to estimation errors (estimating growth parameters) in simulating the ageing effect, thus they partially solve the problem of the templates "representativeness". Moreover, on applying age transformation on real face images, it was observed in [14] that the prediction is good for small age transformations and poor for large age transformation.

Another quite feasible solution is to continuously adapt-enrolled face templates to the variation of the input data available over time [20, 21]. Characteristics of face adaptive biometric systems can be summarized as follows: (1) one no longer needs to collect a large number of biometric samples during the enrolment process; (2) it is no longer necessary to re-enrol or re-train the system (classifier) from the scratch in order to cope up with the changing environment. Recently, template update procedure have received significant boost in biometric community and their efficiency has been proved on evaluation of the resulting performance gain of the system [20, 21]. These methods may be adopted for template improvement or to avoid template ageing.

- *Template improvement*: To increase the representational capability of the enrolled templates by appending new features and samples available during the online operation. Template improvement also includes adaptation to variations like illumination, sensor and other environmental changes causing mismatch conditions due to unrepresentative enrolled templates.
- *Template ageing*: To adapt the templates to the permanent changes of the input data due to the ageing process over time.

Update procedures have been used in template improvement for adaptation to facial variations like expression, lightning, changing sensors and acquisition environment, etc. [22–26]. On the other hand, template update can offer an effective solution to ageing problem as well, for many real-time applications like online banking, ATM, etc., where the interactive of the user is involved and expected after at least certain maximum time period.

However, to the best of our knowledge, in contrast to template improvement, template ageing have not received much attention from adaptation viewpoint. In addition, there exists no study for the comparative performance evaluation of the existing face recognition systems under the ageing effect (temporal variance). This evaluation is important because it will allow to gauge the performance gain of the face recognition systems on employing ageing invariant solutions and facilitate the designer/researcher in choosing the most robust face recognizer to be integrated with ageing invariant solutions for optimal performance. However, none of the available facial ageing databases are specialized only on ageing but other variants, such as illumination, glasses, gender etc., known to influence the performance of the face recognition systems are also present. As a consequence, sole impact of ageing to the performance degradation of the face recognition systems cannot be evaluated. Therefore, it becomes necessary to study the compound effect of ageing with other variate present in the ageing database (covariate analysis). This covariate analysis will allow to analyse the contribution of other variate in the performance degradation of the facial system under the ageing effect. Thus, the four-fold contributions of this chapter are as follows:

1. To evaluate and compare existing face recognition systems based on six different facial representations under the ageing effect.
2. To evaluate the compound effect (covariate analysis) of ageing with other variates such as gender, race, glasses and facial hairs.
3. To introduce template ageing as a concept drift problem.
4. To evaluate the effectiveness of update procedures on facial template ageing for the real-time user interactive scenarios.

The rest of the chapter is organized as follows. Section 6.2 discusses how face goes under ageing. Section 6.3 gives technical detail on template update in terms of attributes, learning methodologies adopted and the similarity between template ageing and concept drift. Section 6.4 reports experimental evaluations and results. Conclusions are finally drawn in Sect. 6.5.

6.2 Face Ageing Process

Face ageing in humans is the result of a multi-dimensional process of physical, psychological, and social change, which affects considerably the appearance of a human face. Both superficial textural wrinkling of the skin and changes in the 3-dimensional (3-D) topography of the underlying structures contribute to ageing of the human face. Ageing-related appearance variation due to bone growth normally occurs throughout childhood and puberty, whereas skin-related effects principally appear in older subjects. From a computer vision perspective, the challenging problem of facial ageing can be described as follows:

1. *Diversity of Ageing Variation*: The rate of ageing and type of age-related effects vary from person to person. Moreover, during different age stages, the facial ageing effects take different forms. Subjects of different ethnicity and genders exhibit typically diverse ageing effects. Outer factors may also lead to diversities in the ageing pattern adopted by different individuals. In addition, anti-ageing products or cosmetic surgeries can also be used to deliberately intercede with the ageing process. Therefore, common ageing patterns or models might not be applied successfully to all subjects.

2. *Shape and Texture*: Generally, facial ageing causes both change in shape and texture of the human face. During formative years, facial shape variations are predominantly manifested, while during later stages of adulthood textural variations in the form of wrinkles and other skin artifacts take precedence over shape variations. Since facial ageing introduces progressive variations in facial appearances, the characterizing models should account for the temporal nature of the induced variations. Therefore, facial ageing may be modelled by attributing facial shape and facial texture as functions of time.

3. *Factors*: Ideally, facial ageing effects are induced by multiple factors, namely *intrinsic* and *extrinsic* factors. Biological elements causing ageing are known as intrinsic factors, whereas environmental influences are called extrinsic factors. Intrinsic factors causing facial ageing are mainly due to the natural changes such as loss in elasticity of facial muscles, bone growth, facial fat tone and volume along with individual's gender, ethnicity, and idiosyncratic features (i.e. features purely unique to the individual such as hyperdynamic facial expressions). Extrinsic factors that may also lead to facial ageing are dietary habits, drug use, lifestyle, use of anti-ageing products/make-ups, cosmetic surgeries, health and psychological and climatic conditions (e.g. skin ageing due to exposure to solar ultraviolet rays known as photoageing.)

4. *Controllability*: Unlike other variates such as illumination, expression, pose and glasses etc., under normal mode the effects of ageing can not be controlled and/or reversed. Thus, it is not possible to rely on the cooperation of the person for eliminating ageing variation during face image capture. In addition, the process of collecting training data suitable for studying the effects of ageing requires long-time intervals.

5. *Feature Selection*: One of the important step of modelling facial ageing or age-invariant FR method is identifying the appropriate form of data that provides a fair description of the process. The forms of the data that might assist characterizing facial growth are fiducial features (2D or 3D) extracted from age-separated faces, 2D facial imagery or 3D facial scans extracted from individuals across different ages, face anthropometric measurements extracted from a population etc. Also, the data could be individual specific or population specific.

6.3 Face Ageing and Template Update

6.3.1 Face Template Update Methods

A typical face recognition system is trained with a limited training data and under static environmental conditions. However, over the time enrolled templates become unrepresentative or outdated to the changing environment and variations in the input data, thus causing degradation in the system's performance. Novel solution to this issue is named 'template update' method, which is continuously adapting enrolled templates to the variation of the input data available during the real-time operation of the system. The continuous template update can handle appropriately the performance loss owing to outdated templates using various techniques to learn the input data variations. The template update procedures can be distinguished on the basis of the following attributes [20, 21]:

- *Learning methodology adopted*: The well-known template update methods are self-training and co-training [21]. In Self-training, the classifier or the system updates itself, whereas in co-training two complementary biometrics update each other to the variation of the input data. Detailed information on different learning methodologies can be found in [21].
- *Online versus Offline*: In online method, the system is updated as soon as the input data is available, while in offline method, after a batch of data has been collected. Online systems are order sensitive in which sequence of instances are observed, one instance at a time, not necessarily in equally spaced time intervals.
- *Supervised or unsupervised*: For supervised methods, input data is labelled by the human supervisor in an offline manner. For unsupervised methods, the input data is labelled by the classifier itself. The positively labelled data, either in supervised or unsupervised mode, is used for adapting the enrolled templates.
- *Template management*: The methods can also be differentiated on the basis of memory buffer utilization, and how the updated template sets are maintained.

  - Appending based: New features/samples are appended to the feature/template set of the old template(s), keeping full memory intact.
  - Replacement based: New template is replaced with the old template. They are not a memory-based methods.

Algorithm 1 The Self-training algorithm

- For each user i:
- Given: enrolled template(s) X consisting of $x_i^{1:t}$ samples.
- Train g1 (a face classifier) using X.
- Loop:
- On availability of the input sample x_i^{t+1}.
- Let g1 to label x_i^{t+1}.
- Use x_i^{t+1} to update the template set if it is positively classified with high confidence. (both appending where $X = x_i^{1:t+1}$ or replacement based $X = x_i^{t+1}$ management schemes can be used in update).
- Re-train g1 using X.

6.3.2 Commonly Adopted Learning Methodologies

The most commonly adopted learning methodology in the state-of-the-art literature [21, 22, 27, 28] is *self-training* [20] based learning, where the classifier adapts itself to the variation of the input data. Self-training methods are consist of two stages. First, label assignment to the input data. Second, adaptation. Generally, these methods are continuous incremental learning techniques, in which the positively classified (with high confidence) input samples are used to update the template gallery. Both the appending or replacement-based template management strategies can be adopted for update purpose. They are also unsupervised user-specific process, i.e. adopted independently for each user [22, 27, 28].

Certain, supervised methods have also been introduced in the literature in which the human supervisor labels the input data [25, 29], which is then used for update. They are offline methods and the update process is carried after certain fixed-time period. Further information about other learning techniques such as co-training or graph-based methods can be found in [20, 21].

The detailed description of the self-training method is given in the Algorithm 1. For supervised methods, the algorithm will be same except that instead of classifier assigning label to the input data, a human supervisor would assign labels.

6.3.3 Face Template Update State-of-the-Art

According to the literature, update mechanism has been customarily adopted for template improvement. Moreover, it has been also reported that there is a significant performance gain using different learning methodologies. For instance, reference [23] evaluates the template update effect on *GEFA* (Gradual evolution of facial appearance) but the database is acquired over a period of five months which fails to capture ageing variations. Table 6.1 refers to the database characteristics of the state-of-the-art update procedures for face recognition systems. The database characteristics contain variations related to mismatch conditions such as change in illumination, sensor, occlusion, pose, expression, etc. We can infer two main aspects from referred works in Table 6.1. First, significant performance gain is observed on the adoption of update procedures. Second, the commonly adopted learning technique is self-training.

Table 6.1 Database characteristics of the state-of-the-art methods for update procedures

References	Database	Variations
[22]	Homemade	Expression, pose and illumination
[23]	GEFA	Expression, illumination, pose and uncontrolled background
[24]	Equinox	Illumination
[25]	BANCA	Expression, illumination and uncontrolled background
[26]	Big brother	Pose and illumination
[30]	AR	Expression, occlusion and illumination

6.3.4 Template Ageing, Update and Concept Drift

In comparison to other facial variations (pose, illumination, etc.), adaptation to template ageing deserves a dedicated treatment of its own, since ageing is a life long process. Ageing also brings gradual changes in the data distribution over time, thus causing performance loss as a result of template becoming outdated. Moreover, order-sensitive adaptation is needed, namely template is adapted using the input samples in the order of its availability.

These factors indicate that template ageing process is very similar to the concept drift theory [7], based on the fact that real-world concepts change with time resulting in underlying data distribution to change. The changes in the data distribution may be incremental or decremental. In other words, these changes may show increasing or decreasing trend. Therefore, causing the model learned on the old concept inconsistent with the new data. One of the efficient solutions as offered for the concept drift problem is regular update.

6.4 Experimental Analysis

6.4.1 Performance Evaluation of Face Recognition Systems Under Ageing Effects

Facial Representations: In this study, following six facial representations were considered: Local Binary Pattern (LBP) [31], Multi-scale Local Binary Patterns (MLBP) [32], LPQ (Local Phase Quantization) [33], LTP (Local Ternary Patterns) [34], EBGM (Elastic Bunch Graph Matching) [35], SIFT (Scale Invariant Feature Transform) [36] and SURF (Speeded Up Robust Features) [37].

Briefly, LBP operator forms labels for the image pixels by thresholding the neighbourhood of each pixel with the centre value and considering the result as a binary number [31]. MLBP is an extended version of LBP using multiple radii and offering the advantage of scale invariance [32]. LPQ utilizes phase information computed locally in a window. The phases of the four low-frequency coefficients are decorrelated and uniformly quantized [33]. In LTP [34], the binary code in LBP are replaced by the ternary code using central pixel value. EBGM [35] localizes a set of landmark features and extracts Gabor jets at landmark positions. SIFT features are efficiently detected through a staged filtering approach and are highly distinctive [36]. SURF relies on integral images for image convolutions (using a Hessian matrix-based measure for the detector and a distribution-based descriptor) [37].

Dataset: We used MORPH [38] dataset comprises of thousand of facial images of individuals across time and collected in real-world conditions (not a controlled collection). This dataset also include essential meta-data, such as age, sex, race, glasses, facial hair, etc. A subset of 631 subjects from MORPH (1700 images) with

Fig. 6.1 Sample images from MORPH databases for a user at 27, 31 and 36th year of age

about 3 images per subject are used in this study. Age range of the subjects are [15, 68]. The characteristics and image samples of MORPH dataset are shown in Table 6.5 and Fig. 6.1, respectively.

Protocol: First of all, facial features are extracted from all images in the database using the considered facial representations. Then, the following steps are performed for performance evaluation:

- Similarity (dissimilarity) matrix is computed using all-pair matching of facial features. Matching scores are divided into an authentic (genuine) and an impostor score distribution.
- The dataset (matching scores) is bootstrapped at the user level, i.e. subset of users are selected with replacement for performance evaluation.
- Performance has been evaluated on calculating area under curve (AUC) statistic on the bootstrapped dataset. AUC is computed as a function of true accept rate (TAR) and false accept rate (FAR) as:

$$AUC = \int_0^1 TAR(FAR)dFAR \qquad (6.1)$$

The AUC value ranges from 0 to 1.
- Variation in AUC on the bootstrapped dataset is recorded as mean $\pm$ std.
- Finally, face recognition systems are ranked in the descending order on the basis of their average AUC on the bootstrapped dataset.

Results: Table 6.2 quotes the AUC values as *mean $\pm$ std* and percentile statistics on the bootstrapped dataset for all the facial representations. These facial representations are mentioned in the descending order on the basis of their AUC values.

It can be seen that all the systems resulted in low performance on the MORPH facial ageing database. MLBP-based facial representation outperformed other facial representations under the ageing effect. These results indicate that MLBP is able to, to some extent, locate discriminative information even under the presence of profound facial ageing between the pair of images. Nevertheless, it could be interesting to integrate MLBP based facial recognition system with age-invariant solutions and gauge the improvement over the baseline performance, as here evaluated to be 0.66 (recorded as AUC) in Table 6.2.

Table 6.2 AUC values obtained on the performance evaluation of six facial representations under the ageing effect on the MORPH database

References	Face representation	Mean ± std	Percentiles [25%, 50%, 75%]
[32]	**MLBP**	0.66 ± 0.02	[0.58, 0.60, 0.64]
[31]	LBP	0.64 ± 0.08	[0.56, 0.59, 0.63]
[33]	LPQ	0.62 ± 0.01	[0.59, 0.60, 0.61]
[35]	EBGM	0.60 ± 0.02	[0.56, 0.57, 0.59]
[34]	LTP	0.55 ± 0.00	[0.50, 0.51, 0.54]
[37]	SURF	0.52 ± 0.01	[0.51, 0.52, 0.53]
[36]	SIFT	0.51 ± 0.08	[0.50, 0.51, 0.51]

6.4.2 Covariate Analysis

The low performance of the considered face recognition systems in Sect. 6.4.1 can not only be solely attributed to the ageing effect, but also to other variates present in the database. Thus, we perform covariate analysis evaluating the contribution of other variates (such as gender, race and glasses) to the performance degradation of the face recognition system under the ageing effect.

Protocol: The aim of this study is to gauge the impact of these covariates under the ageing affect. Therefore, the following covariates are extracted from the MORPH meta-data file. These extracted covariates are explained as follows:

- Age {Young and Old}. Old age is assigned to subjects above 40 years.
- Race {White, Non-white}. Self-explanatory.
- Gender {Male, Female}. Self-explanatory.
- Glasses {Yes, No}. Self-explanatory.
- Facial Hair {Yes, No}. There were many subjects who had thin hairs, beards or not clean shaven.

Using these extracted covariates, single and joint-factor analysis are performed. In single-factor analysis, only one of the available covariates is kept constant and others are allowed to vary. For joint-factor analysis, values of multiple covariates are kept fixed, and the performance of the system is gauged under the ageing impact. These analysis aimed at determining the favorable covariate values under the influence of ageing. The database is broken down into different subsets as follows:

Age {Young (550 subjects) and Old (130 subjects)}
Race {White (171 subjects), Non-white (460 subjects)}
Gender {Male (515 subjects), Female (116 subjects)}
Glasses {Yes (36 subjects), No (612 subjects)}
Facial Hair {Yes (400 subjects), No (342 subjects)}

For the joint-factor analysis, the database is broken down into subsets as mentioned below (compound effect of only those covariates could be analysed for which sufficient number of subjects were obtained)

Non-white and Male (B+M) {373 subjects}
Non-white and Female (B+F) {87 subjects}
White and Male (W+M) {142 subjects}
White and Female (W+F) {29 subjects}
Facial Hair and Glasses (Fh+Gl) {23 subjects}
White and Male (W+M) {142 subjects}
Facial Hair and No Glasses (Fh+NGl){420 subjects}

The bootstrapped version (similar to performance evaluation) of each subset of the database is evaluated using AUC for all the six facial representations.

Results: Results of the single-factor analysis are shown in Table 6.3. It can be seen from the Table 6.3 that MLBP continues to outperform other face representations even for the single-factor analysis.

Table 6.4 shows the results of the joint-factor analysis for the top three good performing facial representations. Most of the results are in accordance with the results of single-factor analysis. For instance, non-white males are easier to recognize than non-white females under the ageing affect (by 8 % for MLBP). Similar observation holds for white males and white females (by about 2 % for MLBP). MLBP usually performed better than other descriptors except for the case of facial hair and no glasses, where LPQ performed better than MLBP by about 5 %.

6.4.3 Template Ageing as a Concept Drift Problem

In this section, we carried out detailed analysis of template ageing as a problem of concept drift.

Dataset: Besides MORPH database, we also used FGNET [17] ageing database containing facial images of number of subjects at different ages. This database has been generated as a part of the European Union project FGNET (Face and Gesture Recognition Research Network).

The characteristics of MORPH and FGNET databases can be found in Table 6.5. Apart from ageing variations, these databases also offer changes in illumination, poses, expression, beard and moustaches, spectacles, hats etc., thus offering a challenging situation. Figure 6.2 shows image samples of an individual from FGNET ageing database.

Protocol: Using FGNET dataset, on the basis of the age of the samples, we followed the notation that each user i consisting of $x_i^{1:T}$ genuine samples in ascending order, i.e. age of x_i^t is lesser than all the other x_i^{t+1} samples. We analysed the genuine pdfs (probability distribution functions) of a randomly chosen user for the FGNET

Table 6.3 AUC values for the single-factor analysis on the bootstrapped version of MORPH dataset

Covariates	Values	MLBP [32]	LBP [31]	LPQ [33]	LTP [34]	EBGM [35]	SURF [37]	SIFT [36]
Age	Young	0.64 ± 0.04	0.62 ± 0.1	0.55 ± 0.03	0.54 ± 0.01	0.54 ± 0.04	0.54 ± 0.02	0.53 ± 0.05
	Old	0.69 ± 0.03	0.67 ± 0.05	0.58 ± 0.03	0.56 ± 0.02	0.56 ± 0.05	0.56 ± 0.03	0.55 ± 0.06
Race	Non-white	0.73 ± 0.01	0.61 ± 0.02	0.54 ± 0.03	0.55 ± 0.01	0.55 ± 0.02	0.55 ± 0.02	0.53 ± 0.05
	White	0.75 ± 0.03	0.62 ± 0.08	0.55 ± 0.03	0.56 ± 0.02	0.56 ± 0.01	0.57 ± 0.03	0.55 ± 0.06
Gender	Male	0.75 ± 0.06	0.62 ± 0.04	0.57 ± 0.04	0.56 ± 0.01	0.55 ± 0.04	0.55 ± 0.01	0.53 ± 0.03
	Female	0.71 ± 0.08	0.61 ± 0.03	0.55 ± 0.03	0.54 ± 0.07	0.54 ± 0.03	0.55 ± 0.04	0.54 ± 0.04
Glasses	Yes	0.72 ± 0.03	0.60 ± 0.03	0.55 ± 0.02	0.55 ± 0.04	0.54 ± 0.03	0.54 ± 0.01	0.51 ± 0.01
	No	0.76 ± 0.06	0.62 ± 0.03	0.56 ± 0.01	0.56 ± 0.03	0.55 ± 0.02	0.55 ± 0.01	0.52 ± 0.02
Facial Hair	Yes	0.77 ± 0.05	0.63 ± 0.03	0.57 ± 0.02	0.55 ± 0.01	0.53 ± 0.01	0.53 ± 0.03	0.52 ± 0.02
	No	0.73 ± 0.04	0.61 ± 0.04	0.55 ± 0.03	0.54 ± 0.01	0.52 ± 0.02	0.52 ± 0.05	0.51 ± 0.04

Table 6.4 AUC values for the Joint-factor analysis on the bootstrapped version of MORPH dataset

Multi-variates	MLBP [32]	LBP [31]	LPQ [33]
B+M	0.65 ± 0.01	0.62 ± 0.03	0.62 ± 0.04
B+F	0.60 ± 0.05	0.60 ± 0.02	0.61 ± 0.02
W+M	0.65 ± 0.07	0.64 ± 0.08	0.63 ± 0.06
W+F	0.64 ± 0.03	0.63 ± 0.04	0.62 ± 0.06
Fh+Gl	0.65 ± 0.01	0.63 ± 0.02	0.62 ± 0.04
Fh+NGl	0.70 ± 0.06	0.65 ± 0.05	0.74 ± 0.05

Table 6.5 The characteristics of the FGNET and MORPH databases as used in the experiments

Characteristics	FGNET	MORPH
No. of subjects	82	631
Average no. of images per subject	6–18	3
Age range	0–69	15–68
Other Intra-class variations		
Illumination	Yes	Yes
Poses	Yes	Yes
Expression	Yes	Yes
Beards and moustaches	Yes	Yes
Spectacles	Yes	Yes
Hats	Yes	Yes

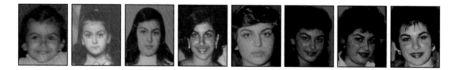

Fig. 6.2 Sample images from FGNET databases for a user at different years (3–41)

database by partitioning the samples into two age groups as shown in Fig. 6.3. The pdf of the first group is obtained by computing the scores within the group, and the pdf of the second group is obtained by computing matching scores via comparing the samples of the first group to that of the second group.

Further, using MORPH database, for each user i, the genuine scores are computed by comparing each sample x_i^j to all the other x_i^k samples, such that age of x_i^j is lesser than all the other x_i^k samples. This same process is repeated for all the N users in the database. Then, all the obtained genuine scores are clustered on the basis of age difference between the two samples used for score computation, irrespective of the user. The clustered scores are then averaged (mean) and mean score versus the age difference is plotted as shown in Fig. 6.4. The x axis in Fig. 6.4 shows the specific age

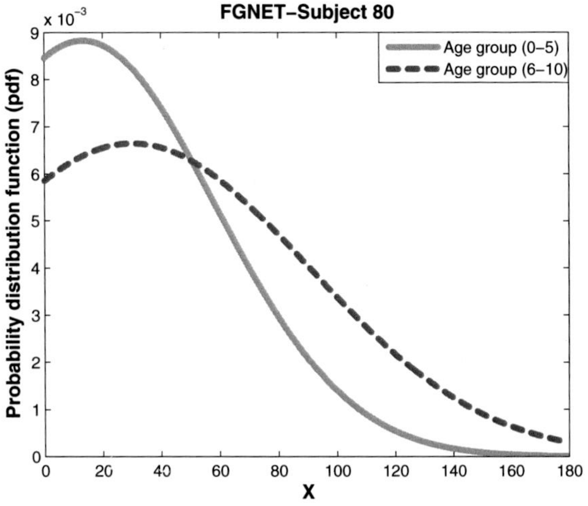

Fig. 6.3 The probability distribution functions (pdfs) drawn by partitioning the image samples into two age groups and computing the genuine score distribution of first group individually. Pdf for the second group is computed by comparing the samples of the first group to the second group. This figure shows the changing distribution the input samples shows ageing variation and learned templates become outdated

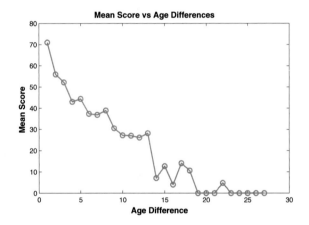

Fig. 6.4 Scores clustered on the basis of age differences between two samples. Mean score for each cluster versus the age difference shows that the presence of decremental drift in the data distribution as a result of time lapse

difference and y axis shows the mean of all the genuine scores belonging to samples with a specific age difference.

Results: Fig. 6.3 shows that the data distribution changes over time as a result of ageing process. While Fig. 6.4 indicates that drift is decremental. The drift is called decremental because the mean of the scores shows decreasing trend as the age differences among the samples increases. Moreover, Fig. 6.4 points out that template update can be an effective technique provided that the new sample is available within certain gap otherwise the updated template will again be outdated to the new incoming samples.

Algorithm 2 Protocol adopted for Join test and adapt strategy

---
- Given: initial gallery set G=($\bigcup g^1_{i=1:N}$), consisting of first image for each person.
- Train face classifier using G
- Loop: For each user i
- Loop j=2:T-1

    - On availability of probe sample x^j_i.
    - Evaluate the performance of the classifier using Rank-1 accuracy.
    - Update the gallery set g_i using the probe sample x^j_i based on supervised and self-training technique.
    - Updated template sets are maintained using both the appending-based and replacement-based update technique.

6.4.4 Evaluation of Template Update Procedures on Facial Ageing

In this section, we perform experimental evaluation of template update for facial ageing. Our experiments used online self-training-based learning technique with supervised and unsupervised label assignment using both, the appending and replacement-based template management strategies. In similar manner, other learning methodologies can be efficiently exploited as well.

Protocol: The effectiveness of the template update for the ageing process can be evaluated using

- how much performance gain is expected?
- the maximum time limit before which the new sample is required for the update procedure to be effective?

Online (incremental) learning updates the system as soon as the input sample is available. For the evaluation of these systems *Join test and adapt*, strategy has been commonly adopted in the literature [21], in which the available input sample is first used to evaluate the performance of the updated system, followed by update [22, 28, 30]. We have also adopted *join test and adapt*-based update and performance evaluation due to its efficiency in better utilizing the limited set of available samples. The complete protocol for online learning together with performance evaluation is given in Algorithm 2, where T is the maximum no. of instances available in the database.

Commercial VeriLook software [39] face identification engine has been used for experimental analysis which consists of two basic modules enrolment and identification. Based on the face identification terminology, during enrolment face image is captured, aligned, face detected and feature sets are extracted and template formed, representing the gallery g_i of a user i. The gallery set G consist of all the gallery images of N subjects in the database, i.e. $G = \cup^N_{i=1} g_i$. On identification, probe image

Table 6.6 (%) Rank-1 accuracy and performance gain of the system that keeps updating in comparison to baseline classifier that does not update

Database	Baseline System (Rank-1 (%) accuracy)	Appending (supervised)	Replacement (supervised)	Appending (unsupervised)
FGNET	12.79	70.09	68.68	29.70
MORPH	66.62	88.83	82.46	82.59

Both the supervised (for appending and replacement-based scheme for managing)and self-training (with appending-based management) learning methodologies are evaluated

p presented to the system is matched with the gallery set G, and the identities are retrieved on the basis of matching score above a set threshold. Performance evaluation of the identification system is done using Rank-1 accuracy at 1 % FAR operating point, which means that % number of times correct identity is ranked first in the list of retrieved identities.

Results: Table 6.6 presents the baseline accuracy (when the first image is in the template gallery for each user) and the averaged Rank-1 accuracy of the system that continuously adapts itself. There is a remarkable increase in the performance of supervised labelling-based self-training for both appending and replacement-based template management schemes. However, for unsupervised label assignment (self-training) using appending-based technique, FGNET does not show much improvement due to the challenging nature of the dataset, thus resulting in very low-genuine scores. It can also be seen that appending-based supervised labelling is better than replacement-based update, i.e. retention of prior knowledge (or old templates) is worthwhile and help to attain more performance gain. Hence, template update schemes can result in substantial performance gain of the system under ageing.

Figures 6.5 and 6.6 show the Rank-1 accuracy obtained on different probe sets for the system that with online update and those obtained for base classifiers for MORPH and FGNET databases, respectively. It can be seen that after every update iteration, better performance is observed on the evaluation of the next probe set.

As mentioned before, another vital question is *what is the maximum time limit before which the new sample is required for the update procedure to be effective?*. Accordingly, the difference in the averaged Rank-1 accuracy of the update-based system with those of the baseline for the users grouped on the basis of averaged age difference among their samples is computed for MORPH database using appending-based supervised labelling technique as given in Table 6.7. The results obtained with FGNET database were qualitatively very similar.

Although no specific trend can be noticed in the performance gain in respect to averaged age difference among the samples. But, it can be seen that till the averaged age difference of nine years, the gain due to update is in double figures then from 10

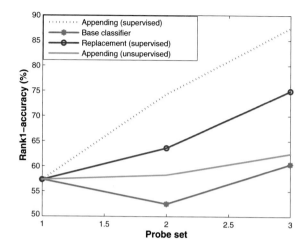

Fig. 6.5 Rank 1-accuracy obtained for different probe sets for online learning-based system in comparison to those obtained for baseline classifier for MORPH database

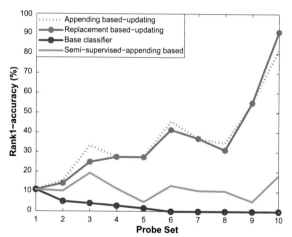

Fig. 6.6 Rank 1-accuracy obtained for different probe sets for online learning-based system in comparison to those obtained for baseline classifier for FGNET database

to 14 years, where gain degrades sharply. It can also be noticed that after 14 years, there is no gain in the performance due to the update procedure. This observation depicts that the maximum time gap in the availability of input sample for the effective update procedure is 9 years according to the Table 6.7. This figure will change with the database, acquisition set-up and the classifier used. Nevertheless, experimental results prove the efficacy of the update procedures for the variations due to template ageing.

Table 6.7 Difference in the averaged Rank-1 accuracy of the update-based system with those of the baseline for the users grouped on the basis of averaged age difference among their samples

Age difference	# No. of subjects	Performance gain (offline) supervised (appending)	Performance gain (online) unsupervised (appending)
1	78	13.02	5.5
2	89	10	9.14
3	63	15.74	3.6
4	49	11.93	7.4
5	56	26	1.97
6	48	20	13.52
7	29	11.59	3.50
8	22	13.89	4.16
9	22	27.38	0
10	13	4.54	0
11	15	1.78	0
12	10	0	0
13	8	6.25	0
14	6	8.3	0
15	4	0	0
16	7	0	0
17	4	0	0
18	4	0	0
19	1	0	0
21	2	0	0
22	2	0	0

6.5 Conclusions

In the long term, the performance of a face recognition system is affected by the face ageing, which causes significant alterations in the human faces. Therefore, in this study, first six baseline facial representations based on local features were evaluated under facial ageing impact, to show that how severely ageing degrades the performances. Further, the compound effect of ageing with other variate (such as gender, race, glasses and facial hair) are systematically analysed. The covariate analysis results were mostly in accordance with the results of ageing factor analysis. Performance loss due to facial ageing is the result of change in the data distribution causing the templates to be outdated over time, making thus the process similar to the concept drift theory. The solution to this problem is age transformation-based techniques for ageing and de-ageing solutions using the models trained on age-separated images, which is not quite feasible for real-time evaluation. In addition, these methods are prone to estimation errors in simulating the ageing effect. Recently update procedures have been introduced offering effective and simple way for the template adaptation. However, effectiveness of these methods for ageing variations have not been evaluated till date. Experimental results, in this chapter, on commonly used facial ageing databases (FGNET and MORPH) ensures that template update can effectively adapt the system to temporal variance.

To conclude, it is worth mentioning that the field of facial template update has yet not fully developed for the real-time implementation. This is due to the number of open issues associated with this field [20]. One of the major problems is the impostor intrusion into the updated template set due to successful zero and non-zero effort impostor attacks.

Acknowledgments   The authors would like to thank Dr. Ajita Rattani of the Department of Computer Science and Engineering, Michigan State University (USA) for her valuable suggestions.

References

1. Poh, N., Kittler, J., Rattani, A., Tistarelli, M.: Group-specific score normalization for biometric systems. In: Proceedings of the IEEE Conference on Computer Vision and Pattern Recognition Workshops (CVPRW), pp. 38–45 (2010)
2. Akhtar, Z.: Security of multimodal biometric systems against spoof attacks. PhD thesis, University of Cagliari, Italy (2012)
3. Akhtar, Z., Micheloni, C., Piciarelli, C., Foresti, G.L.: MoBio_LivDet: mobile biometric liveness detection. In: IEEE International Conference on Advanced Video and Signal based Surveillance (AVSS), pp. 187–192 (2014)
4. Akhtar, Z., Kale, S., Alfarid, N.: Spoof attacks on multimodal biometric systems. In: Proceedings of the International Conference on Information and Network Technology (ICINT), pp. 46–51 (2011)
5. Akhtar, Z., Alfarid, N.: Secure learning algorithm for multimodal biometric systems against spoof attacks. Proceedings on the International Conference on Information and Network Technology (ICINT), pp. 52–57 (2011)
6. FRVT, http://www.nist.gov/itl/iad/ig/frvt-2013.cfm/ (2013)
7. Tsymbal, A.: The problem of concept drift: Definitions and related work. Department of Computer Science, Trinity College, Ireland (2004)
8. Akhtar, Z., Ahmed, A., Erdem, C.E., Foresti, G.L.: Biometric template update under facial aging. In: Proceedings of the IEEE Symposium on Computational Intelligence in Biometrics and Identity Management (IEEE SSCI-CIBIM) (2014)
9. Akhtar, Z., Rattani, A., Hadid, A., Tistarelli, M.: Face recognition under ageing effect: a comparative analysis. In: Proceedings of the International Conference on Image Analysis and Processing (ICIAP), pp. 309–318 (2013)
10. Flynn, P.J., Bowyer, K.W., Phillips, P.J.: Assessment of time dependency in face recognition: an initial study. In: Proceedings of the 4th International Conference on Audio and Video based Biometric Person Authentication, pp. 44–51 (2003)
11. Ling, H., Soatto, S. Ramanathan, N., Jacobs, D.W.: A study of face recognition as people age. In: Proceedings of the 11th IEEE International Conference on Computer Vision (ICCV), pp. 1-8 (2007)
12. Rattani, A., Freni, B., Marcialis, G.L., Roli, F.: Template update methods in adaptive biometric systems: a critical review. In: Proceedings of the International Conference on Biometrics (ICB), pp. 847–857 (2009)

13. Lanitis, A., Taylor, C.J., Cootes, T.F.: Toward automatic simulation of aging effects on face images. IEEE Trans. Pattern Anal. Mach. Intell. **24**(4), 442–455 (2002)
14. Ramanathan, N., Chellappa, R.: Face verification across age progression. In: Proceedings of the IEEE Conference on Computer Vision and Pattern Recognition (CVPR), pp. 462–469 (2005)
15. Ramanathan, N., Chellappa, R.: Modeling age progression in young faces. In: Proceedings of the IEEE Conference on Computer Vision and Pattern Recognition (CVPR), pp. 387–394 (2006)
16. Park, U., Tong, Y., Jain, A.K.: Age-invariant face recognition. IEEE Trans. Pattern Anal. Mach. Intell. **32**(5), 947–954 (2010)
17. FGNET Aging Database: http://www.fgnet.rsunit.com/
18. Ricanek, K.J., Tesafaye, T.: Morph: a longitudinal image database of normal adult age-progression. In: Proceedings of the International Conference on Automatic Face and Gesture Recognition (FG), pp. 341–345 (2006)
19. Nixon, N., Galassi, P.: The Brown Sisters, Thirty-three Years. In The Museum of Modern Art, NY (2007)
20. Rattani, A.: Adaptive biometric system based on template update procedures. PhD thesis, University of Cagliari, Italy (2010)
21. Poh, N., Rattani, A., Roli, F.: Critical analysis of adaptive biometric systems. IET Biometrics **1**(4), 179–187 (2012)
22. Liu, X., Chen, T., Thornton, S.M.: Eigenspace updating for non-stationary process and its application to face recognition. Pattern Recogn. **36**, 1945–1959 (2003)
23. Pavani, S.K., Sukno, F.M., Butakoff, C., Planes, X., Frangi, A.F.: A confidence based update rule for self-updating human face recognition systems. In: Proceedings of the International Conference on Biometrics (ICB), pp. 151–160 (2009)
24. Rattani, A., Marcialis, G.L., Roli, F.: Biometric template update using the graph mincut: a case study in face verification. In: Proceedings of 6th IEEE Biometric Symposium (2008)
25. Poh, N., Kittler, J., Marcel, S., Matrouf, D., Bonastre, J.F.: Model and score adaptation for biometric systems: coping with device interoperability and changing acquisition conditions. In: Proceedings 20th International Conference on Pattern Recognition (ICPR), pp. 1229–1232 (2010)
26. Franco, A., Maio, D., Maltoni, D.: Incremental template updating for face recognition in home environments. Pattern Recogn. **43**, 2891–2903 (2010)
27. Jiang, X., Ser, W.: Online fingerprint template improvement. IEEE Trans. PAMI **8**, 1121–1126 (2002)
28. Ryu, C., Hakil, K., Jain, A.: Template adaptation based fingerprint verification. In: Proceedings of the International Conference on Pattern Recognition (ICPR), pp. 582–585 (2006)
29. Uludag, U., Ross, A., Jain, A.: Biometric template selection and update: a case study in finger-prints. Pattern Recogn. **37**(7), 1533–1542 (2004)
30. Roli, F., Marcialis, G.L.: Semi-supervised pca-based face recognition using self training. Pro-ceedings of the Joint IAPR International Workshops on S+SSPR (2006)
31. Ahonen, T., Hadid, A., Pietikainen, M.: Face description with local binary patterns: application to face recognition. IEEE Trans. Pattern Anal. Mach. Intell. **28**(12), 2037–2041 (2006)
32. Chan, C.-H., Kittler, J., Messer, K.: Multi-scale local binary pattern histograms for face recog-nition. In: ICB, pp. 809–818 (2007)
33. Ahonen, T., Rahtu, E., Ojansivu, V., Heikkil, J.: Recognition of blurred faces using local phase quantization. In: Proceedings of the International Conference on Pattern Recognition, pp. 8–11 (2008)
34. Tan, X., Triggs, B.: Enhanced local texture feature sets for face recognition under difficult lighting conditions. IEEE Trans. Image Process. **19**(6), 1635–1650 (2010)
35. Wiskott, L., Fellous, J.M., Kruger, N., Malsburg, C.: Face recognition by elastic bunch graph matching. IEEE Trans. PAMI **19**(7), 775–780 (1997)
36. Kisku, D.R., Rattani, A., Grosso, E., Tistarelli, M.: Face identification by sift-based complete graph topology. In: Proceedings of 5th IEEE International Workshop on Automatic Identifica-tion Advanced Technologies, pp. 63–68 (2007)

37. Dreuw, P., Steingrube, P., Hanselmann, H., Ney, H.: SURF-Face: face recognition under view-point consistency constraints. In: Proceedings BMVC, pp. 1–11 (2009)
38. MORPH database: http://www.faceaginggroup.com/projects-morph.html/
39. Verilook: http://www.neurotechnology.com/

Chapter 7
An Adaptive Score Level Fusion Scheme for Multimodal Biometric Systems

Kamlesh Tiwari and Phalguni Gupta

Abstract This chapter presents a score level fusion scheme for multimodal biometric system. There multiple scores corresponding to the matchings of different biometric samples are fused for taking decision on similarity. Proposed score normalization is an threshold alignment and range compression scheme. It utilizes statistical properties of the score distribution. The proposed scheme has been tested over a multimodal database which is constructed using three publicly available database. Experimental results have shown the significant performance boost.

7.1 Introduction

Biometrics provides a very intuitive way to recognize human by his physiological or behavioural characteristics. Various traits like face, palmprint, fingerprint, vein pattern, iris, knuckle, voice, gait, etc., have been explored and found useful [6]. Although some of these traits are found to be fairly accurate, but it is hard to design a system which provide 100 % accuracy using only a single trait. This is because of behaviour of user and quality of acquired sample. Accuracy and reliability of the biometric system can be improved if one uses more than one trait. Such systems are called as multimodal systems.

Sources of multiple biometric evidences can be classified in six categories as proposed in [15], and these are multi-sensor, multi-algorithm, multi-instance, multi-sample, multimodal, and hybrid systems. Multi-sensor system uses more than one sensor to acquires the same biometric sample, while multi-algorithm system applies more than one feature extraction and matching technique to obtain matching score and multi-instance system acquire more than one available variation of same trait-like

K. Tiwari (✉)
Indian Institute of Technology Kanpur, Kanpur 208016, India
e-mail: ktiwari@cse.iitk.ac.in

P. Gupta
National Institute of Technical Teachers' Training and Research, Kolkata 700106, India
e-mail: pg@iitk.ac.in

© Springer International Publishing Switzerland 2015
A. Rattani et al. (eds.), *Adaptive Biometric Systems*,
Advances in Computer Vision and Pattern Recognition,
DOI 10.1007/978-3-319-24865-3_7

another finger or another iris. Multi-sample system uses more than one sample one by one using same sensor. Multimodal system involves ore than one traits. Further, a hybrid system uses combination of these five (Fig. 7.1).

A multimodal system makes use of multiple biometric traits to obtain a fused score. Fusion can be done at are (1) feature extraction level, (2) matching score level or (3) decision level [14]. Out of them, matching score level fusion is the most promising because it can give more freedom to consider the best-suited feature extraction and matching techniques of individual trait. Score level fusion becomes tricky because of the fact that the range of matching scores produced by matchers may be different, and they may follow different distributions. Therefore, score normalization becomes very important for fusion. There exist various score normalization techniques for multimodal systems such as min–max, decimal-scaling, z-score, median and median absolute deviation (MAD), double sigmoid, *tanh* and bi-weight localization etc. [7]. These normalization techniques are found to be sensitive to outliers and user-specific weights assignment to normalized scores can perform better.

A framework is proposed in [9] for optimal combination of multimodal match scores which is based on the likelihood ratio test. A dynamic reconciliation scheme for score fusion is proposed in [18]. Normalization scheme proposed in [5] is derived from min/max and using three biometric traits fingerprint, face and vein pattern has shown that it could attain better performance than [9]. A fusion scheme for complementary biometric modalities using face and palmprint which is proposed in [13] uses Log-Gabor transformations and particle swarm optimization. Relevance

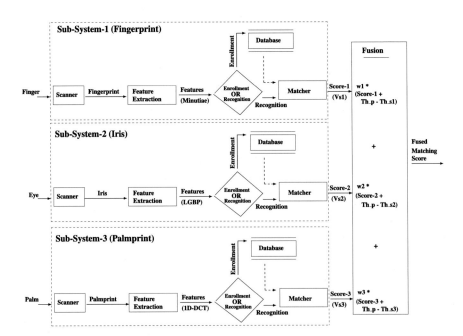

Fig. 7.1 Block diagram of the proposed scheme for score normalization and fusion

Vector Machine is used in [8], and a ranking-based user-specific fusion strategy is proposed in [12]. There is a need for a sophisticated fusion scheme which is adaptive to the underline statistical properties of a biometric system such as error rate and accuracy for their normalization and fusion.

This chapter describes an efficient score normalization scheme for a multimodal biometric system. Relative fusion weights for individual traits are assigned by utilizing their individual performance parameters. A multimodal database involving 100 subjects from public databases *viz*. FVC2006-DB2-A, CASIS-V4-Lamp and PolyU has been constructed, and experimental results are shown on them. Rest of the chapter is organized as follows. Section 7.2 describes the proposed normalization strategy. Section 7.3 elaborates the experimental setup, database used and results obtained. Conclusions are presented in the last section.

7.2 Proposed Normalization Scheme

This section proposes a score level normalization and fusion scheme which applies threshold alignment and range compression. This particular fusion scenario appears when a person presents more than one biometric sample for verification (possibly from different traits; say a fingerprint and a iris). It uses the person's identification to retrieve the stored template from the database and subsequently matches the respective biometric samples. Matching on any unimodal biometric system produces its similarity/dissimilarity score. These scores are normalized to fuse and to produce a final matching score.

The score level fusion is tricky as the range of matching score produced by the two matchers may be different, and they may also follow different distributions. Also there are situations when performance of the different matchers are priori known, and different weights are required to be assigned to different matching. A normalization scheme which preserves the relative monotonicity of original score in corresponding normalized scheme is called as order preserving score normalization scheme (OPSNS). Any OPSNS ensures that for two given matching scores v_1 and v_2 with $v_1 < v_2$, their normalization score also satisfies the same relationship $normalized(v_1) < normalized(v_2)$. It has also a very interesting property. When it is uniformly applied on all the matching scores, they do not produce any effect to the performance of biometric system. This is because of the fact that the performance of the system is evaluated based on a particular threshold which decides the number of false accept and false reject cases. Applying OPSNS can vary the decision threshold, but it retains relative ordering of the scores. Therefore, it does not affect the number of false accept and false reject cases and as a result, performance of the system remains unchanged. There exist several OPSNS such as min–max, z-score, etc.

Matching scores produced by a matcher are not randomly distributed in its range. Also they do not follow normal distribution. But individually they are expected to obey the normal distribution (Gaussian) [6]. Both genuine and imposter distributions have different statistical properties. Ideally these two distributions need to be well separated but in practical, they have overlaps. A decision threshold is chosen such that it separates the two distributions and achieves maximum possible accuracy. Weighing of the matching scores is necessary when the involved systems have different recognition accuracies to restrict the adverse effect of low confidence system. The normalization strategy discussed in this chapter applies threshold alignment followed by weighting to obtain the final matching score.

7.2.1 Threshold Alignment

Let the decision thresholds used by n unimodal biometric systems $S_1, S_2, \ldots, S_n$ be $Th_{S_1}, Th_{S_2}, \ldots, Th_{S_n}$, respectively. The system designer arbitrarily chooses a single and fixed pivot threshold Th_p which is used for threshold alignment of all unimodal biometric matching scores. Threshold aligned score v'_{S_i} for a min–max-normalized matching score v_{S_i} of S_i can be obtained using following formula.

$$v'_{S_i} = v_{S_i} + (Th_p - Th_{S_i}) \tag{7.1}$$

In our experiment, we have utilized three unimodal biometric systems say S_1, S_2 and S_3. Let their decision thresholds be Th_{S_1}, Th_{S_2} and Th_{S_3}, respectively. Threshold aligned score v'_{S_1} for a min–max-normalized matching score v_{S_1} of S_1 is obtained by

$$v'_{S_1} = v_{S_1} + (Th_p - Th_{S_1}) \tag{7.2}$$

Similarly, the threshold aligned scores v'_{S_2} and v'_{S_3} for v_{S_2} and v_{S_3} of S_2 and S_3 can be obtained by

$$v'_{S_2} = v_{S_2} + (Th_p - Th_{S_2}) \tag{7.3}$$

$$v'_{S_3} = v_{S_3} + (Th_p - Th_{S_3}) \tag{7.4}$$

Purpose of the initial min–max normalization is to bring the scores of different modalities in same range. This step helps to align appropriate thresholds.

7.2.2 Score Weighting

Relative multiplicative weights of the three systems are obtained with the help of equal error rate (EER) of the unimodal biometric system. Since EER is inversely related to the accuracy; a polynomial of the form $c_1 \times (EER)^{-2} + c_2 \times (EER)^{-1} + c_3$ is used to obtain relative fusion weights where c_1, c_2, c_3 are constants. The value of constants are empirically determined using regression on small subset of the database. Let EER_{S_i} be the equal error rate of an individual biometric sub-system S_i then its relative multiplicative weight wr_{s_i} is determined as below.

$$wr_{S_i} = c_1 \times (EER_{S_i})^{-2} + c_2 \times (EER_{S_i})^{-1} + c_3 \qquad (7.5)$$

Let three individual biometric systems S_1, S_2 and S_3 have their equal error rates as EER_{S_1}, EER_{S_2} and EER_{S_3}, respectively. Then their relative multiplicative weights wr_{S_1}, wr_{S_2} and wr_{S_3} are determined as below.

$$wr_{S_1} = c_1 \times (EER_{S_1})^{-2} + c_2 \times EER_{S_1}^{-1} + c_3 \qquad (7.6)$$

$$wr_{S_2} = c_1 \times (EER_{S_2})^{-2} + c_2 \times EER_{S_2}^{-1} + c_3 \qquad (7.7)$$

$$wr_{S_3} = c_1 \times (EER_{S_3})^{-2} + c_2 \times EER_{S_3}^{-1} + c_3 \qquad (7.8)$$

Absolute weight w_i of an individual biometric system S_i is evaluated with the help of relative multiplicative weights of all participating individual biometric subsystems $w_1, w_2, \ldots, w_n$ by

$$w_i = \frac{wr_{S_i}}{\sum_{k=1}^{n} wr_{S_k}} \qquad (7.9)$$

The fused score $v_{S_1 S_2 S_3, \ldots, S_n}$ for the scores of n biometric traits is obtained by combining $v'_{S_1}, v'_{S_2}, \ldots, v'_{S_n}$ and $w_1, w_2, \ldots, w_n$ as below.

$$v_{S_1 S_2 S_3, \ldots, S_n} = w_1 \times v'_{S_1} + w_2 \times v'_{S_2} + \cdots + w_n \times v'_{S_n} \qquad (7.10)$$

7.3 Results

The proposed scheme has been tested on a multimodal database comprising of three biometric traits *viz.* fingerprint, iris and palmprint. Biometric samples of each individual trait has been taken from publicly available database. Fingerprints are considered from FVC2006-DB2-A, while irises are from CASIA-V4-Lamp and palmprints from PolyU database. Since the number of subjects and number of biometric samples per subject differ across the databases, we have selected 100 subjects and three samples from each of the databases for the construction of multimodal database. Selection is anticipated to contain difficult to recognize users of goat and lamb category [16]. One out of the three images per subject per trait is used for training and remaining two are used for testing. Training and testing sets are mutually exclusive, and there is no overlapping subject in training and testing set.

(a) FVC2006-DB2-A Database [2]. This database contains fingerprints of 140 subjects. For every subject, there are 12 fingerprint images. In all, there are 1680 images acquired through optical sensor of size 400×560 and 569 dpi resolution. A sample image of FVC2006-DB2-A is shown in Fig. 7.2a.

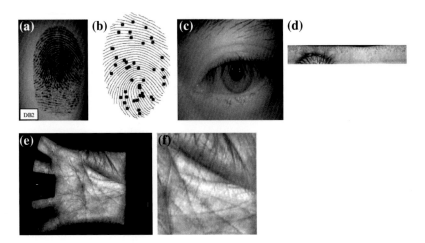

Fig. 7.2 Example of typical biometric samples of fingerprint, iris and palmprint along with their minutiae, normalization and ROI. **a** Fingerprint of FVC2006-DB2-A, **b** minutiae in a fingerprint image, **c** iris of CASIA-V4-Lamp, **d** normalized iris image, **e** palmprint of PolyU and **f** extracted palmprint ROI

(b) CASIA-V4-Lamp Database [3]. This database consists of 16,212 images collected from 411 subjects having 819 distinct irises and 20 images per iris. Iris images contain non-linear deformation due to variation of visible illumination. A sample iris image from CASIA-V4-Lamp database and its corresponding normalized iris strip are shown in Fig. 7.2c, d respectively. Normalized iris image is cartesian to polar transformation of the iris portion of eye image.

(c) PolyU Database [17]. This database contains of 7,752 grayscale images from 193 users corresponding to 386 different palms. Around 17 images for each palm are collected in two sessions. Images are acquired placing pegs and using CCD at spatial resolution of 75 dpi and 256 gray levels. A sample hand images from PolyU database and its corresponding extracted palmprint ROI are shown in Fig. 7.2e, f respectively.

Performance of a typical biometric system is analysed with respect to its error rates in rejecting a genuine or accepting an imposer based on the statistical properties of the matching score distribution. Standard performance metrics are described below.

False Accept Rate (FAR) refers to the rate at which unauthorized individuals (imposers) are accepted by the biometric system as a valid user. False acceptance is an error which signifies the probability of an intrusion. Its value should be as low as possible.

False Reject Rate (FRR) refers to the probability that a biometric system fails to identify a genuine subject. False rejection is an error which signifies the denial of access to an authentic user of the system. Its value should also be as low as possible.

Equal Error Rate (EER) is the rate at where FAR and FRR are equal. EER is a typical choice of system operation because it balances the user inconvenience and security. Lower the value of EER better is the system.

Correct Recognition Rate (CRR) represents rank-1 accuracy and signifies the percentage of the best match which pertains to the same subject. It can be measured as

$$CRR = (N_1/N_2) \times 100$$

where N_1 is the number of correct (Non-False) recognition and N_2 is the total number of images in the testing set. Higher the CRR better is the system.

Decidability Index (DI) signifies how well the genuine and imposter matching scores are separated. DI is defined by

$$DI = \frac{|\mu_g - \mu_i|}{\sqrt{(\sigma_g^2 - \sigma_i^2)/2}}$$

where μ_g and μ_i are the mean and σ_g^2 and σ_i^2 are the variances of genuine and imposter distributions, respectively. Decidability index is found to be high for highly accurate systems.

7.3.1 Experimental Results

Minutiae of fingerprints of FVC2006-DB2-A are extracted by *mindtct* and matching is done using *bozorth3* [11]. Figure 7.2b shows typical fingerprint with minutiae marked on it. The unimodal fingerprint system has achieved CRR of 96 % and EER of 4.1 % on the subset of FVC2006-DB2-A database containing three fingerprints of 100 users. Histogram of normalized dissimilarity matching scores obtained through *bozorth3* is plotted in Fig. 7.3a.

Eye images of CASIA-V4-Lamp database are subjected to iris segmentation using an improved circular hough transform and robust integro-differential operator [4] that detects the inner and the outer iris boundary. The segmented iris is normalized to polar coordinates to get an image like Fig. 7.2d. It is further preprocessed using LGBP (Local Gradient Binary Pattern) and the corners features are extracted and matched using dissimilarity measure CIOF (Corners having Inconsistent Optical Flow) [10]. The iris-based unimodal biometric system has achieved an CRR of 92 % and EER of 11.52 % on the selected subset of CASIA-V4-Lamp database. Histogram of normalized dissimilarity matching scores through CIOF is plotted in Fig. 7.3b.

Palm images of PolyU database are segmented using the method proposed in [1] to extract a square region of interest (ROI) as shown in Fig. 7.2f. ROI is further enhanced and its features are extracted. Matching technique [1] on the selected subset of PolyU database has achieved the CRR of 90 % and EER of 12.52 %. Genuine and imposter score histograms of normalized dissimilarity values are presented in Fig. 7.3c.

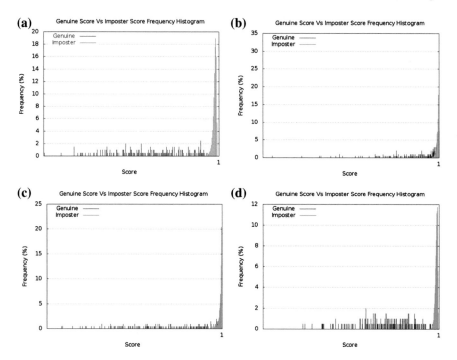

Fig. 7.3 Fused genuine and imposter score histograms. **a** Histogram of normalized fingerprint scores, **b** histogram of normalized iris scores, **c** histogram of normalized palmprint scores and **d** histogram of proposed normalization scheme on multimodal database

Operating characteristics of the three individual biometrics systems provide working threshold, which are used to align the scores. EER of the three unimodal biometric systems have been used to determine the relative weights. The proposed normalization-based multimodal system has achieved an CRR of 99 % and EER of 0.99 % on the selected database. Histogram of the normalized score is plotted in Fig. 7.3d. ROC curves of three individual unimodal biometric systems of fingerprint, iris and palm are plotted in Fig. 7.4. It can be seen that the performance of the multimodal system is better than any unimodal biometric system.

A comparison with score normalization and fusion strategies like minimum, maximum, majority voting, median, summation, product, tanh normalization, median and MAD, double Sigmoid and z-score has been presented in Table 7.1. It has been observed that CRR of the multimodal system is 94 % and EER is 11.13 % when maximum score among the three matchers is considered as fused score. It has falsely rejected 23 genuine out of 200 and falsely accepted 2130 imposters out of 19800. Median has slightly better rank-1 accuracy and low error rate as compared to maximum. Performance of majority voting fusion strategy is found to be better than maximum. Although it is evident from Table 7.1 that summation and product fusion

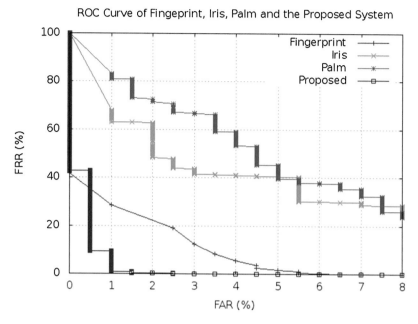

Fig. 7.4 Receiver operating characteristic *curve* for fingerprint, iris, palm and proposed scheme

strategies outperforms majority voting in our experiment. Both strategies have rank-1 accuracy of 99 % and EER of 2.00 % but DI of *product* fusion strategy is better.

It can also be observed that non-linear fusion strategy like tanh has CRR of 99 % and EER of 1.99 %. Another strategy, called Median and MAD, has been found to be performing similar to tanh strategy with slight improvement in EER to 1.90 %. Fusion strategy of double sigmoid has further reduced the error rate to 1.50 % but its DI is found to be 0.71 which is low. It can be observed that DI of z-score normalization is better than double sigmoid, minimum and tanh. Minimum EER value that has been achieved across all other competing normalization strategies is 1.50 %. The proposed fusion strategy has achieved rank-1 accuracy (CRR) of 99 % with a DI of 1.81, and its equal error rate (EER) is found to be 0.99 % which is a significant improvement over other techniques. The proposed strategy has the lowest number of false rejection (2 out of 200) and false acceptance (186 out of 19800). ROC curves of score normalization and fusion strategies are plotted in Fig. 7.5a. It clearly indicates the superiority of the proposed system.

Histogram of genuine and imposter score of the proposed fusion scheme is shown in Fig. 7.3d. The best genuine and the best imposter scores for a particular query image can be determined by considering the lowest matching score among all genuine and the highest matching score among all imposters. Figure 7.5b shows the best genuine and imposter scores for all the query image of the proposed normalization scheme. It can be seen that there is a good separation between average genuine and imposter matching scores.

Table 7.1 Values of CRR, ERR and DI for different settings

	CRR %	EER %	DI	FR	FA
System on individual trait					
Fingerprint	96	4.10	1.60	9	492
Iris	92	11.50	0.80	23	2276
Palmprint	90	12.50	1.13	25	2474
Fusion strategy of multimodal system					
Maximum	94	11.13	0.90	23	2130
Median	96	5.64	1.43	11	845
Majority voting	98	2.50	1.30	5	494
Summation	99	2.00	1.76	4	395
Product	99	2.00	1.91	4	395
tanh and summation	99	1.99	1.58	4	394
Median and MAD and summation	99	1.90	1.76	4	359
Double sigmoid and summation	99	1.50	0.71	3	294
Minimum	99	1.50	1.64	3	296
z-score and summation	99	1.50	1.75	3	296
Proposed	99	0.99	1.81	2	186

Falsely reject (FR) and falsely accept (FA) are out of 200 genuine and 19800 imposer matchings, respectively

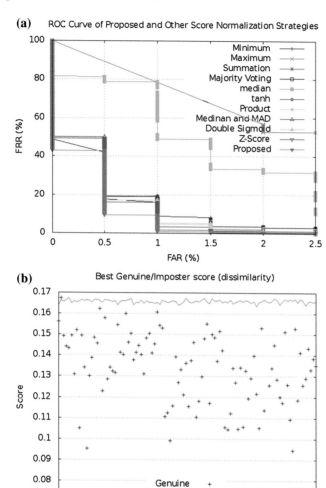

Fig. 7.5 ROC *curves* of various fusion schemes and best genuine and imposter score distribution. **a** Receiver operating characteristic proposed and other fusion schemes. **b** Fused genuine and imposter score histogram

7.4 Conclusions

This chapter discusses an efficient score level fusion strategy. It is suitable for score fusion of multimodal biometrics system. It performs range compression and threshold alignment for score normalization which makes use of statistical properties of biometric score distribution. The fusion scheme is tested on a multimodal biometric database which is constructed with the help of publicly available biometric database of fingerprint FVC2006, iris database CASIA-V4-Lamp and palmprint database PoluU. Comparison of the proposed normalization scheme with other fusion strategies like maximum, majority voting, median, product, tanh, median and MAD, double Sigmoid and z-score etc. have been studied. Experimental results have shown the superiority of the system which has achieved significant decrease in error rates.

References

1. Badrinath, G.S., Tiwari, K., Gupta, P.: An efficient palmprint based recognition system using 1D-DCT features. In: Intelligent Computing Technology. LNCS, vol. 7389, pp. 594–601. Springer, Heidelberg (2012)
2. Cappelli, R., Ferrara, M., Franco, A., Maltoni, D.: Fingerprint verification competition 2006. Biometric Technol. Today 15(7), 7–9 (2007)
3. CASIA-IrisV4 database. http://www.cbsr.ia.ac.cn/china/Iris%20Databases%20CH.asp
4. Daugman, J.: How iris recognition works. IEEE Trans. Circ. Syst. Video Technol. 14(1), 21–30 (2004)
5. He, M., Horng, S.-J., Fan, P., Run, R.-S., Chen, R.-J., Lai, J.-L., Khan, M.K., Sentosa, K.O.: Performance evaluation of score level fusion in multimodal biometric systems. Pattern Recognit. 43(5), 1789–1800 (2010)
6. Jain, A.K., Flynn, P.J., Ross, A.A.: Handbook of Biometrics. Springer (2008)
7. Jain, A., Nandakumar, K., Ross, A.: Score normalization in multimodal biometric systems. Pattern Recognit
8. Mehrotra, H., Vatsa, M., Singh, R., Banshidhar, M.: Biometric match score fusion using rvm: a case study in multi-unit iris recognition. In: Computer Society Conference on Computer Vision and Pattern Recognition Workshops, pp. 65–70. IEEE (2012)
9. Nandakumar, K., Chen, Y., Sarat C.D., Jain, A.K.: Likelihood ratio-based biometric score fusion. IEEE Trans. Pattern Anal. Mach. Intell. 30(2), 342–347 (2008)
10. Nigam, A., Gupta, P.: Iris recognition using consistent corner optical flow. In: Asian Conference on Computer Vision (ACCV), pp. 358–369. Springer (2013)
11. NIST biometric image software. http://www.nist.gov/itl/iad/ig/nbis.cfm
12. Poh, N., Ross, A., Lee, W., Kittler, J.: A user-specific and selective multimodal biometric fusion strategy by ranking subjects. Pattern Recognit. 46(12), 3341–3357 (2013)
13. Raghavendra, R., Dorizzi, B., Rao, A., Kumar, G.H.: Designing efficient fusion schemes for multimodal biometric systems using face and palmprint. Pattern Recognit. 44(5), 1076–1088 (2011)
14. Ross, A., Jain, A.: Information fusion in biometrics. Pattern Recognit. Lett. 24(13), 2115–2125 (2003)
15. Ross, A.A., Jain, A.K., Nandakumar, K.: Information fusion in biometrics. In: Handbook of Multibiometrics, pp. 37–58 (2006)
16. Ross, A., Rattani, A., Tistarelli, M.: Exploiting the doddington zoo effect in biometric fusion. In: International Conference on Biometrics, pp. 1–7. IEEE (2009)

17. The PolyU palmprint database. http://www.comp.polyu.edu.hk/biometrics
18. Vatsa, M., Singh, R., Noore, A., Ross, A.: On the dynamic selection of biometric fusion algorithms. IEEE Trans. Inf. Forensics Secur. 5(3), 470–479 (2010)

Index

© Springer International Publishing Switzerland 2015
A. Rattani et al. (eds.), *Adaptive Biometric Systems*,
Advances in Computer Vision and Pattern Recognition,
DOI 10.1007/978-3-319-24865-3

Printed in the United States
By Bookmasters